Planning and Managing Scientific Research

Planning and Managing Scientific Research

A guide for the beginning researcher

Brian Kennett

Australian National University

PRESS

ANU PRESS

Published by ANU Press
The Australian National University
Canberra ACT 0200, Australia
Email: anupress@anu.edu.au
This title is also available online at http://press.anu.edu.au

---

National Library of Australia Cataloguing-in-Publication entry

Author: Kennett, B. L. N. (Brian Leslie Norman), 1948- author

Title: Planning and Managing Scientific Research :
                   A guide for the beginning researcher / B.L.N. Kennett.

ISBN: 9781925021585 (paperback)   9781925021592 (ebook)

Subjects: Research--Management.
                   Technological innovations--Management.
                   Universities and colleges--Research work.

Dewey Number: 658.57

Cover design by Nic Welbourn and layout by ANU Press

Contents

Contents

Preface

This book has been developed and expanded from materials prepared for a course delivered to Masters students in Earth Science at the Australian National University. These students had a broad range of experience and discipline base, and so from the beginning the treatment was rather general. The aim is to build an understanding of the nature of scientific research, and the way in which it can be planned and managed. The emphasis is on broadly applicable principles that can be of value irrespective of discipline, derived from my extensive research, editorial and management experience.

This short work is aimed at beginning researchers, particularly in the later stages of Ph.D. work and Postdoctoral workers, but should also be of value for independent researchers. In the early days of research guidance is likely from research supervisors and advisors. But, all too soon, a researcher can confront the situation where they are expected to be able to develop their own projects and make them work. This book is intended to help this transition, and to provide a framework that should be of value into the future. Each researcher will establish their own style and mode of work, my aim is to aid them to make this process as effective as possible.

At a number of points in the book I have made use of material prepared by different organisations, such as guidelines for proposals. Such material is designated by distinctive type, and often appears as boxed text. Apart from this material, the content is my own and I have prepared the figures and charts specifically for this book.

I am grateful for the feedback provided by the students who participated in the various deliveries of the course. Their input helped significantly in clarifying the issues confronting the novice researcher. In the course there was a strong interactive component, which is difficult to reproduce outside the class environment. Some self-help exercises are suggested based on these interactions. These exercises are likely to be of greatest value when discussed with others.

Brian Kennett
Canberra, March 2014.

Chapter 1
The Nature of Research and Innovation

1.1 Introduction

Scientific research encompasses many different styles of activity. The nature of the work varies significantly, depending on the problem and the discipline. The work practices in laboratory-based experimental studies are distinctly different from field-based observations, or theoretical investigations. Nevertheless, broad principles can be recognised that unify the approach to goal-driven research. It is the aim of this short book to provide a guide to successful practice in research, drawing on experience of major research projects and research management.

An important first step is to understand the context of a research project, and the classes of circumstances and limitations that are likely to be associated with a particular class of activity. Projects do not exist in isolation, so it is useful to look across the full range of scientific endeavour because analogies and lessons can often be drawn from other fields.

In this first chapter we look at the nature of scientific research and innovation in a general way. It is useful to develop the skill to look at a project from the outside, and so attempt to see it in a broader context. We examine the nature of styles of research and the classes of external constraints and the way that they influence the nature of the enterprise.

The second chapter is concerned with the full life cycle of a project from concepts, through planning to execution. We draw on both research and management principles. We examine the construction and assessment of project proposals, and the practical problems to be faced once a project is under way. Project development (either formal or informal) lies at the heart of research, and presents many challenges — in managing the process, the people involved and the reporting requirements. Forethought helps greatly with project management and, in the third chapter, I provide insight into project planning and tracking based on successful projects.

An important part of research activity is the effective communication of results, so that work is known and linked to the broader research scene. In the fourth chapter we consider modes of communication from preparing a paper for publication, delivering seminars or conference presentations and establishing an internet presence for a project.

The fifth chapter is directed to a range of issues that arise from the nature of research and the necessary human interactions. We consider relations within a research team and, more broadly, recognising incipient problems and more fundamental ethical issues.

The sixth chapter presents two case studies, based on the author's experience, to illustrate the need for dynamic management and adaptability to achieve a successful outcome. In reality few projects proceed entirely as expected, and some planning for contingencies can markedly improve outcomes. Changes along the way can be positive, for example, when new methods become available, rather than just negative, as when the external environment changes. In any case, adaptation is the key to resilience and long term success.

1.2 What is research?

Scientific research is characterised by a systematic approach to the generation of new knowledge, building on previous work yet subjecting it to close scrutiny to determine any failings. Irrespective of the type of research activity (theoretical, experimental, observational) the results have to be substantiated, and the work must be reproducible. The continual process of testing and crosschecking from many different quarters leads to a consensus on the major aspects of scientific fields.

The presence of this consensus position provides a good basis for the broad advance of a field, but can provide a barrier to the acceptance of new concepts, results or methods that lie outside the current framework. It can take persistence to deflect the path of a field, but once a new framework has been accepted it becomes the new norm. Thomas Kuhn has termed this schema of gradual evolution, punctuated with major shifts in viewpoint, a 'paradigm shift'. The basis of observational and experimental results remains unchanged, but is now interpreted in new ways. Commonly it is the advent of new classes of result, which are discordant with the existing framework, or are only representable in a clumsy way, that forces ultimate change. The Copernican revolution to a heliocentric solar system, the development of quantum theory, and the adoption of plate tectonics in the Earth sciences represent three such major shifts in conceptual framework. The decoding of the nature of DNA has radically changed the nature of biological research, and modified the emphasis of the research endeavour leading to a shift in focus.

Yet, the new standpoint is not immutable. Our conventional structures are best regarded as working hypotheses under continual test. When discrepant results arise, and are validated by independent work, the early steps generally involve some patching of the structure, for example, by introducing additional complications. Then, finally, a new concept emerges to explain the full suite of results in a more elegant way. It is not given to most researchers to effect major shifts in their fields, but nevertheless results can have unexpected significance.

Current science represents a large and complex enterprise, and research has acquired a varied apparatus of funding schemes and expectations. I hope that the advice in this book will help researchers to recognise that many aspects of the research environment are common to all fields. Good planning and organisation cannot replace scientific insight, but will help to maximise research returns.

Building new research results into innovative products and services is undertaken in a somewhat different way than unconstrained research. The funding sources tend to come with stronger expectations and greater restrictions. I will therefore try to indicate the differences that arise in different circumstances, whilst emphasising generally applicable principles.

1.3 Styles of research and innovation

In many contexts in research we are required to classify the nature of the work being done, such as at the time of submission of a proposal to a funding agency. Such information is also frequently collected for statistical purposes. As we shall see, it is not entirely straightforward to classify the nature of research, and the pattern of research activity can change over the course of a project.

Many researchers spend their career in individual or small-group activities. Yet, there is an increasing tendency towards larger collaborative ventures that cross disciplinary boundaries. Larger programs bring with them the need for more explicit organisation and management, which tends to impose more constraints on constituent projects.

1.3.1 Categories of research

The first type of situation in which a researcher will come into contact with classifying research is usually in submission of a paper for publication. A preliminary sorting in relation to the scientific discipline base will have occurred with the choice of the journal, but then the authors are expected to provide specific information on sub-fields that is employed in the review process. Similar considerations enter in the submission of grant proposals.

Discipline and objectives

Research activities can be classified by scientific discipline, or the goal of the work such as mineral exploration. Indeed both types of information are complementary and help to place the nature of the work in context.

A wide range of classification systems are used in different countries

Table 1.1 Field of Research Codes for Earth Sciences (ANZSRC 2008)

04	**Earth Sciences**
0401	Atmospheric Sciences
0402	Geochemistry
0403	Geology
0404	Geophysics
0405	Oceanography
0406	Physical Geography and Environmental Geoscience
040601	*Geomorphology and Regolith and Landscape Evolution*
040602	*Glaciology*
040603	*Hydrogeology*
040604	*Natural Hazards*
040605	*Palaeoclimatology*
040606	*Quaternary Environments*
040607	*Surface Processes*
040608	*Surfacewater Hydrology*
0409	Other Earth Sciences

across the world. There are similarities in style, but the way in which disciplines are recognised and the choice of discipline groupings vary considerably. Classification schemes designed originally for statistical purposes are often employed to specify the character of research in grant proposals. A short extract from the Australian and New Zealand Standard Research Classification (ANZSRC), 2008 *Fields of Research Codes* for the Earth Sciences is shown in Table 1. An attempt is made to break down the main topic into a set of specified sub-fields, with a final entry to capture what has otherwise been missed. Rather than being confined to a single sub-field, it is usually possible to provide percentage of effort distributed across a number of codes.

These classification schemes for research activity tend to remain in place for many years, and so do not readily take into account the evolution of a field; nor are they generally suited to interdisciplinary work. Revision is not simple, even where extensive consultation is undertaken, and often simply shifts the location of problems. Even so it is usually possible to achieve a reasonable alignment on discipline specifiers.

There tend to be more problems with classifiers directed at the areas that will benefit from the research, since these tend to be oriented towards specific applications. Such *socio-economic objectives* have taken on greater significance in recent years as many countries see a major role for public investment in research as aiming for wealth generation in the national interest. The classifications for economic activity are frequently finely graded for industrial applications with many entries. Yet, in many cases, a pure research component is relegated to a subsidiary category assigned to a broad field, for example, Physical Sciences.

Such categorisation of research may seem esoteric when first encountered, but is frequently required for a project in formal grant applications. The specification of discipline, or of proportions between disciplines, needs to be considered carefully since it may well play a role in the assignment of assessors for projects.

Character of research

A further class of information that is frequently sought about research projects concerns the character of the research being undertaken. This is often presented in terms of choices from three broad classes of activity: *basic research*, *strategic basic research* and *applied research*. Even though these terms are regularly used there is little uniformity in definitions; in part because the terms may be regarded as self explanatory.

We can indicate the character of these three research categories as follows:

- *Basic* – curiosity driven though often tied to understanding a broad class of problem; sometimes termed 'blue sky?' research as, for example, the search for the unification of gravitation and the other physical forces in a single theory.
- *Strategic Basic* – research directed towards a clearly defined goal, can frequently include a strong component of new knowledge.
- *Applied* – research directed at a specific topic with the intention of immediate application, generally builds on earlier more fundamental work, but can often include the exploitation of existing knowledge in new ways

The boundaries between these three categories are blurred, and an individual project may contain components of all three. The balance between the different aspects and styles of research may also change as a project evolves.

The three-stage classification of
Basic
Strategic Basic
Applied
is too coarse to represent the nuances of many projects. A five-point scale with intermediate states is probably more representative of the gradations encountered between projects.

Alternatively one can think about assigning a three-stage classification to different aspects of the work:
Concept
Research in progress
Delivery of results
and then derive an aggregate result.

The perception of those working on a project and outside observers will frequently be different. Often the insider perceives a *basic* component that is not as evident from outside. In the author's experience the *strategic basic* element is a significant component in many projects, even if not explicitly stated.

A description of research character is frequently sought in relation to statistical information on the way in which national research effort is directed. The broad classes can also be linked to *socio-economic objectives*: as the research takes on a more strategic or applied character it becomes easier to assign components of the research to specific categories.

Exercise 1-1:
Collect information on a group of research projects from your own or a cognate institution. Analyse the projects in terms of the discipline classifications appropriate to your research environment, and also assign the style of research.
If possible, compare your classifications with those assigned by members of the particular project. Where do you differ and why?

1.3.2 *Types of research activities*

The way in which research is undertaken depends strongly on the style of research activity. In particular the organisation of the work and interactions with other researchers are affected by the nature and scale of the funding source. Across the globe there is a tendency to encourage large collaborative *programs* with a number of subsidiary projects. The increase in size brings with it both organisational and financial issues that require explicit management. In smaller projects informal management can be adequate.

Many researchers are unhappy with the concept of *research management*, since they wish to concentrate solely on the research component, and see management as purely associated with administrative chores. In fact most people employ informal management techniques when they make decisions about where to put in their next effort on a project. As we shall see, even modest projects can benefit from clear planning and tracking using simple tools.

Individual and collaborative research

This is the traditional model of individual, or small group, research activity, which on a larger scale merges into projects built by voluntary collaboration of disciplinary specialists. The projects are typically funded by individual or collaborative investigator grants from a national research agency or equivalent. Research projects can span any of the three research categories; the main differences will arise in the project definition, rather than in the way the project is carried out.

Such grant-supported projects usually have short-term funding (e.g., 2–3 years) and modest reporting requirements. Often projects are pushing at the financial boundaries to achieve their goals, so that monetary pressures can be significant.

At this level much of the management of the project is linked to the process of gaining adequate funding, and establishing good linkages with collaborators. The projects are initiated by the investigators, and to secure funding have clearly developed goals, but little, or no, formal structure.

Multi / interdisciplinary projects

Once projects begin to transcend discipline boundaries, more explicit effort is required to encourage the necessary levels of collaboration between researchers with different backgrounds. Such research activities tend to focus on strategic basic and applied issues, and are most effective where they build on disciplinary strength. Commonly, several investigators will be involved bringing together their expertise to tackle this class of broader issue. Real effort is then needed to set up the necessary interactions between the components.

At an early stage in such projects, an important step is bringing the disparate group of researchers together to establish good communications and a clear understanding of the roles of different participants. Interaction between researchers then needs to be sustained, so that disciplinary strengths can be translated into an effective synergy that bridges the disciplinary boundaries.

Such a complex project can benefit from clear research management, with coordination of financial issues between the multiple partners. There also needs to be an effective path for information flow so that the strands of work can move together.

Many such projects build on data or information from prior results, and there is likely to be a heavy dependence on the quality of *metadata*, the description of the content and nature of data and how it was acquired.

Such *metadata* become critical in many areas that exploit large databases. Considerable work is needed to ensure that all the information necessary for full exploitation in the future is indeed provided. Unfortunately this component of work is rarely given high priority, and has a tendency to be the first to feel the impact of budget shortfalls. The consequences may not be felt until much later.

As emphasis is placed on scrutiny of contentious results, and questions of reproducibility become important, the need for effective *metadata* increases. In the policy arena assertions of suitable courses of action are no longer adequate, and need to be accompanied by explanations of how the advice has been derived and the results on which it is based. These may include experimental results or large-scale computations. In each case sufficient information needs to be present so that, in principle at least, the research can be repeated independently.

Large research programs

Large research programs are now funded by many agencies, and tend to have longer duration and comprise multiple research projects. The reporting requirements and financial control are more complex, so that there is a need for a structure that achieves coordination between the projects. The best results are generally obtained where the projects can influence the overall direction, and there is good communication between projects. For such large programs it is important to establish a cooperative mode of operation between the different component projects, though some inter-project rivalry can be healthy.

A major difficulty in the management of large programs is keeping all projects working on the same time frames. Often, continuing funding is dependent on meeting key *milestones* and *performance indicators*, and this can only be achieved if the various projects provide adequate reporting. Clear advance planning and well-thought out funding agreements at the beginning of the program are important. Many large research programs use independent boards to provide oversight and advice to management. It is important that the relative roles of the program management and the board are clearly defined from the outset, since otherwise much effort can be expended in unnecessary conflict.

The character of a large program depends on the nature and style of the funding source. In some cases the large program is built from the bottom up, by assembling a set of strands promoted by individual investigators. Even then, once the entire program is in view, gaps or needs for linkage may be identified that require directed projects. In other cases the broad outlines of a program are agreed by the participating institutions. Once the funding has been secured, the organisational structure of themes and projects is established. In this case individual investigators have less direct input to the nature of projects, but there are stronger external expectations with regard to project outcomes.

Commissioned research or innovation

Most work in this category falls into the applied field, but sometimes the starting point is sufficiently far from the specific application that it lies in the strategic basic domain. Since the work is commissioned, there is less freedom of action. Considerable care is need with the agreed project definition so that the desired outcomes are feasible. Difficulties can arise where the client wishes to get an answer to a specific question that may represent the symptom of a problem rather than a topic that is directly amenable to attack. In such a case thought needs to be given to examining the problem to see if the project definition can be recast so it is more feasible.

The client will frequently have a fixed budget, which may not be well matched to project needs. This constraint means that project costing has to be done carefully, recognising that people costs may prove a limiting factor. It is unwise to take on a project where expectations and budget do not match.

For commissioned research, extra effort is needed to develop a research plan that satisfies the client's needs, and yet allows sufficient flexibility to allow for development in the light of experience.

Industrial research

Research in an industrial context is closely linked to the economic fortunes of the entity, and can be subject to rapid change or closure if developments have made a project unprofitable, even if the science component is proceeding well. A typical enterprise will have a mix of tactical work (immediate problem-solving), and longer term evolutionary or revolutionary research. The strategic work is the class that may involve external commissioned work or collaborative projects, such as those funded through the Australian Research Council (ARC) Linkage Program.

A strong feature of industrial research is an emphasis on delivery of agreed milestones, and designated review points to assess progress against goals and financial viability. An emphasis of rapid delivery of results means that research teams are expected to be flexible and interdisciplinary. This may require team members to acquire new disciplinary skill sets in a short time frame.

The different cultures in the industrial and academic environments can lead to somewhat divergent views of the same activity in commissioned research. It therefore pays to spend some time on developing a mutual understanding before agreeing to take on such projects.

1.4 Limitations on research activities

Many projects can be self-contained, but few start from scratch. Commonly the project definition will have been built on previously published results and depend, in part, on earlier data, experimental procedures or theoretical frameworks. Such dependencies need to be recognised. Deviations from expectations during a project do not necessarily imply that a project is going wrong, but do need to be recognised and monitored.

1.4.1 Response to external influences

The most obvious case where external influences play a critical role is in large research programs with conditional funding. Apparently burdensome reporting requirements at the project level to the central administration are frequently linked to those on the program as a whole. Coordination of many projects requires information transfer in a more extensive and detailed manner than would satisfy a project's internal needs. Similar issues arise in collaborative projects, since the information level has to be sufficient for full understanding across all participants.

Large programs may also bring with them the requirement to work with particular facilities. In this case, the issue of coordinating such tasks as sample preparation with the availability of equipment can present challenges. Access timetables need to be clearly established and well publicised. Even within a single institution, competition for access time between multiple activities can produce problems and tensions of a similar nature.

A different type of dependencies on factors external to a project arises when there is a need to exploit prior data or samples. There is always the possibility of some class of contamination due to poor collection or analysis procedures. This makes the role of the associated *metadata* critical; yet often this is precisely what is missing. Even within a single research group it can be difficult to reconstruct metadata if it was not recorded in an appropriate form at the time the work was carried out. Notebooks can get lost, likewise old computer files.

Another insidious influence can come from previously collected data and analyses based on a set of assumptions that are no longer adequate. A particular difficulty arises when the original information is sifted, so that components not concordant with a specific model have been suppressed. Such outliers may well contain extra information that allows a model to be challenged.

It always needs to be recognised that there are major differences in research between the exploitation of existing data and the collection of new, specific data. Although work is required to assemble prior information in a form that can exploited for new analysis, this process is usually quicker than collecting new results. It is easy to underestimate the effort needed to set up a new laboratory program or fieldwork campaign. Thus, though new results can meet the strictest protocols, there will be a period before they become available.

On occasion it can be advantageous to consider *outsourcing* some aspects of the work associated with a project, particularly some routine activities. Often there are cost benefits in using a specialised agent, but thereby some control is lost. In particular the timeliness of delivery of results can be an issue. Planning needs to take this into account. It is not good to have a critical dependence on outsourcing, even if there are financial levers on those undertaking the work.

Where a number of investigators are collaborating, it is important to maintain communication so that the different strands of the work can proceed in an effective fashion. Frequently some component of the project will depend on an element undertaken by someone else. The consequences if a component takes longer than expected can be minimised if participants are aware of the delay, particularly if critical dependencies have been identified at the planning stage.

The nature of research is such that, unless a project is a small modification of prior work, some degree of uncertainty and uneven progress will be present. Some researchers prefer to work simultaneously on two, or more, projects so that they can progress one while working out how to overcome bottlenecks on another. Quite apart from the dilution of effort applied to each project, juggling the demands of multiple projects creates its own problems. For example, multiple reporting requirements can fall due at the same time leading to awkward deadlines.

1.4.2 Innovation within boundaries

As we have seen there are many circumstances in which aspects of a research project are constrained. The obvious case is commissioned research, but the constituent projects of a larger research program have to fit in that structure. In such circumstances it is important to understand the specific environment and structure within which the research project sits. For commissioned research there will be pressure for delivery, and it is therefore important to avoid overstating research potential at the outset. It is much better to over-deliver than under-deliver. Much depends on the framing of the questions to be addressed, and the mutual understanding of the parties involved.

In all cases where there are constraints a clear project design and time framework are essential. In particular internally designated *milestones* can be used as a constructive tool. These should be linked to any externally imposed requirements, but do not have to be visible outside the project itself.

1.5 Developing a research topic

A critical component of research is the setting up of a new research topic. Even when this builds on previous work it is worthwhile to spend some time looking at the issues afresh, since the external environment may have changed.

Early in a research career it is likely that the main lines of the topic and the mode of attack will be specified, or at least suggested, by a research supervisor. Nevertheless, it is worthwhile to develop the skills of looking at the project in the light of cognate work in the field.

For a new project the first step is the definition of the questions to be addressed, and an idea of the resources that are needed to make it happen. With this in mind it is then possible to build up a specification of the topic in a way that can form the basis of a proposal, either informal as a means of guiding the subsequent research, or formal as in submission to a funding agency. In the following chapter we will discuss the steps once the topic is defined.

Much of the development of a research topic can appear to grow organically: an idea is discussed and developed to the point where it becomes sufficiently coherent for the outlines of a research concept to emerge. Alternatively, a call for proposals or the submission deadline for a funding agency may stimulate the need for a well-defined research topic conforming to particular requirements.

The set of ideas below are drawn from my experience in developing a wide range of research topics over the years. When only one piece of research is in train at a time the process can be fairly loose, but once a number of stands of research are under way at the same time, it is helpful to have each well specified. In this way interdependencies become clear, and basic documentation exists that can be helpful in recruiting the necessary collaborators for larger projects or certain types of funding schemes.

1.5.1 Framing research questions

A fundamental component of success in research depends on being able to pose a suitable set of questions from which a project can grow. This is not a simple process since relatively innocuous issues may prove to be difficult to pursue with success.

A group of topics that need to be considered in the framing of a project include:

- What do you want to know?
- Incremental ideas or new direction?

- Interesting to yourself and others?
- Important in its own right or in combination with other work?
- How much work is required?

It may seem trite to ask 'what do you need to know', but frequently the difference between a very successful project and a less successful one comes down to the way that the issue is addressed. For example, if there is a specific problem the obvious approach is to tackle this issue directly. However, it may well be that the most successful way is to take a more fundamental line, and search for the root cause of which the problem is a symptom. Such situations are not uncommon in commissioned research, but are not confined to this area.

The pattern of research activity is somewhat different when one can build directly on previous work, compared with striking out a new direction. Where one can exploit earlier relevant studies the initial stages can move more rapidly, but progress may flag when new issues are encountered.

It is difficult to codify the process of innovation. There is no universal formula for success in research. As noted by Louis Pasteur: *in the field of observation chance favours the prepared mind*. Results that do not fit with preconceptions are likely to be more important than those that simply corroborate expectations. Similar considerations occur in theoretical work. A chance association may trigger new insights, and open or reopen a line of research.

The rewards are generally higher when aiming for innovation, but the likelihood of a smooth flow is less. In many cases one may realise that the full importance of a piece of work can only be achieved if cognate projects are also undertaken. The issue then becomes whether it is appropriate to tackle them sequentially, with an extended timescale, or to enlist collaborators for a parallel effort.

At the outset, it can be hard to get a good grasp on how much work will be needed, and commonly this will be underestimated. This component is an important part of building up the full description of a project. Experience helps in all aspects of topic selection, but is not an infallible guide!

1.5.2 Building a topic description

Key question and mode of attack

A critical first step is a clear definition of the target of the research. This may involve testing a hypothesis based on theory or earlier experimental results, or, in more observationally oriented fields, building on conjectures to try to establish a hypothesis. The view of the scientific method developed by Karl

Popper is that science only proceeds by 'falsifying hypotheses', and this is a commonly adopted viewpoint. Yet, such a representation of scientific endeavour provides little place for the creativity needed to even propose hypotheses.

When writing research proposals it may be necessary to frame the work in terms of hypothesis testing. Inevitably many promising ideas will prove wrong, yet many still have some degree of utility. Remember that, in normal conditions, Newton's Laws provide an adequate description of physical behaviour, even though we know they are incomplete. Nevertheless the full panoply of general relativity is needed to get maximum accuracy from Global Navigation Satellite Systems such as GPS.

Once the research target is established, the style of approach to the research issues has to be worked out. Will the research attack the problem directly? Or is it sensible to make a more indirect approach, for example, by tackling a more general issue and then extracting a specific result? There is no general answer, the choice depends on the nature of the problem and the style of work required. A direct attack is often more suitable to laboratory-based experimental work. Indirect modes are common in the observational sciences, but also may be appropriate in a theoretical project where the problem is broken into a sequence of sub-problems.

A further important point to consider is whether the necessary expertise is available to carry out the project as envisaged. If not, can it be acquired using currently available resources or does it need to be brought in by engaging with collaborators?

Research landscape

Before developing a research topic too far, it is useful to understand where it fits in the general research landscape. Important questions are:

- How does the proposed work tie into current thinking?
- What are its dependencies?
- Who are natural collaborators?

There are many sources of information, including recent publications and preprints. Often meetings and conferences prove fertile ground for recognising promising lines of activity.

A good knowledge of the research landscape of your field will also allow you to recognise likely competition and their strengths and weaknesses. This enables you to see what different style can be brought to bear on the research target. The whole process can be informal, but warrants some attention.

Constraints and opportunities

It is important to be able to recognise any constraints that apply to a topic, and also any opportunities that it may provide. Progress is likely to be most effective if the topic can be broken up into stages of research effort of increasing levels of complexity. Thus we need to ask if it is possible to simplify the class of questions being addressed. If so, what is lost? Does the simplified problem contain the essence of the original.

Sometimes, particularly in theoretical work, it can be more effective to seek an approximate solution to a complex problem, rather than a more complete solution to a simplified problem. However, in the effort to isolate the influence of one aspect of a system in experimental work, it is possible to miss important trade-offs between different parameters.

Whatever form such simplification takes, it is important to understand the assumptions that underlie the approach, and the consequent limitations that they may impose. If your work will depend on methods developed by others, make sure that you are aware of any approximations or restrictions involved, and the conditions under which they are valid.

It is always dangerous to have strong dependence on 'black box' components in research, for example, in the use of software for data analysis. Sometimes this is inevitable, as when proprietary material is used, but it is important to understand what is going on, rather than just use it and hope all will be well.

General issues

It takes time to develop a suitable topic to the point where it represents a viable research project. Make sure that you allow sufficient time to digest the current state of the field. There is nothing more frustrating than finding that a research project has been pre-empted by other work. Of course, others may have the same types of ideas since science is a competitive endeavour, and ideas tend to have their season. Fortunately it is rare for projects to be exactly the same, and a good knowledge of the research background will help avoid duplication of effort.

In the process of selecting and developing research topics it can be helpful to seek advice relating to a research target. You will get the most value from such advice if you understand the concepts involved, and what they imply, as well as what is possible. This will mean that you can ask appropriate questions. The better prepared that you are, the better placed you will be to evaluate advice. Even good suggestions may not prove to be useful, since they may represent a different viewpoint on the problem than you have decided to take. It is easy to acquire too many opinions and be hard pressed to know which way to turn.

Be realistic in your expectations, particularly about the time frame needed to make progress. Initial estimates tend to be optimistic, even when based on considerable experience.

> **Exercise 1-2:**
> Characterise your current research topic in terms of: key questions, potential interested parties, likely competitors, expected outcomes.

1.5.3 Research resources

The range of resources available to a researcher has substantially expanded from the traditional materials of printed journals and books. In particular, the electronic versions of journals now frequently include supplementary materials expanding the basic content, such as movies from simulations. Such extra information can often provide greater insight into the thinking behind the results presented in the paper itself. Once a research topic has been determined it is not difficult to arrange e-mail notifications from selected journals for relevant materials as they appear. Some care may be needed in the nature of the specifications employed, since it is possible to be overwhelmed by too thorough notifications.

For most topics there will be a modest number of journals in which the majority of useful articles will appear, but it has always been difficult to catch all relevant materials since some will be published in less obvious locations. In some fields there are long-established abstracting services that have carried a traditional mode of operation into the digital world. For example, the *GeoRef* database in Earth Sciences provides extensive historical information in an accessible form, yet also has a latency of only a month or two for new material. Such resources allow systematic searching using multiple criteria in combination, though it can take some practice to use the search procedures to maximum effect. In many areas of Physics, extensive use of preprints is made ahead of formal publication, and the *arXiv* facility is extensively used, both for storage and searching for new material.

A number of web resources provide access to searches of databases of scientific articles based on criteria such as the presence of a designated word in the title, the name of the author or some more complex composite criterion. The major resources of this type are:

◇ *Web of Science* (Thomson Reuters),
◇ *Scopus* (Elsevier),
◇ *Google Scholar*.

The commercial resources *Web of Science*, and *Scopus* require an institutional subscription, and provide sophisticated filtering by year of publication and general subject, together with information that can help to separate authors with similar names. Along with the details of the publications and potential modes of access, this class of resources provides information on the citation patterns of the publications both with respect to the papers they have cited, and to later papers that cite the publication. It can take some time before material is registered in these databases, particularly when they come from rival publishers.

Citation analysis is well developed for a wide range of scientific journals, and also some conference proceedings, but is more haphazard for books or articles in edited books. *Google Scholar* tends to catch more book-related material, but is not complete. It is advantageous to seek advice on the most effective ways of implementing searches in such web resources, since well framed searches are the most likely to access relevant information.

In addition to the formal scholarly structures, useful information can often be extracted from more general web-based resources. It is important, however, to treat such information with caution since there is no guarantee of reliability or accuracy in material posted on the web. In many cases a search with a standard web search engine, such as *Google*, will return useful information, particularly if the combination of key words employed can narrow the possibilities. Nevertheless, you may find that something of interest lies well down in the precedence order created by the web search engine, so do not rely on the first page of results.

Wikipedia offers a remarkable range of articles on general and scientific topics, with the possibility of updates or corrections from users. Wikipedia is a multi-lingual resource, and entries on the same topic can vary considerably in nature and emphasis depending on the language in which the contribution is provided. I found *Wikipedia* helpful when thinking about the section of the book on project management since pointers were provided to material that I would otherwise have missed.

The websites of institutions and researchers can also provide useful information. Particularly at the individual researcher level, maintenance of sites can be somewhat lax, and so recent information may be missing. Information on research practices on websites will not normally have undergone any peer scrutiny, and so should not be relied on without independent corroboration.

The use of social media in science tends to focus on drawing attention to events or products from projects, but can provide a convenient means of keeping multiple participants in a large project informed. A recent development is *ResearchGate*, which provides access to researcher profiles and their publications, as well as hosting question-and-answer sessions across a broad range of topics. Searches can be carried out by name or keyword. As on a personal website, materials are user provided and so may need to be treated with caution. Such social products are likely to take on even greater significance in the future.

You are well advised to consult your local librarian for information on resources relevant to your particular interests. Frequently, they will be aware of materials that will not necessarily appear in simple searches. Networks of librarians are effective in sharing information about available resources.

Exercise 1-3:
Develop a resource list for your field of interest in terms of: journals, databases and other materials.

Chapter 2
The Life Cycle of a Project

Once one has a clear vision for a research topic, the next step is to develop a project, planning the outlines of the work that will be needed to reach the desired goals and outcomes. Another class of research-related project for which planning is critical is in the development of research infrastructure, such as the commissioning of major equipment.

Normally it will be necessary to seek funding to support the project, and this is done through the submission of a formal research proposal. The requirements for such proposals depend, to some extent, on the individual funding agencies. Nevertheless we can recognise some broad principles that help to provide a good representation of the research concept. Even where a full formal proposal is not required, it is useful to build an informal proposal to capture the features of a project in a succinct form.

Once funding has been secured, detailed planning for the project needs to start in earnest, taking into account the resources available over the time frame of funding. If, for example, new staff have to be recruited, it helps to have made the necessary preparations so that the human resources processes can be set in train as soon as funding is available.

There are well-developed tools to aid project management, and these can be beneficial in a research context. It takes some effort to learn how to use the tools and to set them up, but the return is a much clearer understanding of dependencies, and the critical points of interaction between the different components of a project.

As a research project progresses there are likely to be requirements for reporting on progress through, for example, achievement of milestones and the use of key performance indicators. Although such performance measures are most common for large research programs, the requirements can be imposed on constituent projects. When first encountered such performance measures tend to be regarded as a burdensome nuisance, but can ultimately be a significant help even for a single project.

A simple aspect of project management is tracking the financial situation against the available budget; this is important, but represents only part of what can be achieved. Experience with analysing a project and its development can be of major value when one comes to consider the next research topic.

2.1 Developing a new project

What then needs to be done to develop a new project? We can recognise a number of steps and stages of planning. Initially the process tends to be rather informal, but once funding needs to be sought then more explicit materials are required:

- What is the aim?
  At the beginning it is important to get a clear enunciation of the research target. This will form the base on which you will build the rest of the project plan.
- What are the expectations?
  With a clear target in mind, you can turn to defining appropriate research outcomes associated with the achievement of the target. This will often stimulate thoughts about related outputs, such as publications.
- Is the specification correct?
  The initial thinking should now be sufficient to provide a mental picture of the nature of the research project. It is useful at this point to review the scope of the project:
  ◇ Is the topic well defined and self-contained or is it open-ended?
  ◇ Can you refine your target to produce a clearer vision?
     If the project is relatively open-ended, it may be possible to define a stage marker, for example, a result that is required before any further steps can be taken, and then plan towards that sub-goal. Often only a small change in the definition of target will produce a simplification of the approach, and a consequent gain in terms of ability to achieve the goals. It helps to be flexible in thinking at this early stage, so that you avoid being locked into an ultimately untenable position.
- Establish position in research landscape
  Before you progress too far it is useful to understand where your proposed project sits in the general research landscape:
  ◇ Are you seeking to establish a new direction?
  ◇ Does the project require resources or results from other sources?
     The response to these types of questions will dictate what sort of background material is needed on recent work pertinent to the field. The relevant work may already be well known to you, but it does not hurt to check.
- Earliest version of project plan
  With a general view of the research topic in mind, this is the time to sketch out the nature of the research activity. At this stage the 'plan' may be just a set of notes to summarise the general situation, addressing the main points:
  ◇ What do I need to know to achieve the research goals?

◇ What resources are needed?

◇ What are the appropriate funding sources?

◇ An idea of constraints and bottlenecks.

- Create a potential time line
  This is a good point to try to sketch out the main stages of the research and the way that they interact. You need to take into account the availability of personnel, and any delays likely if recruitment is required. Also, equipment procurement will have a significant lead time that can impinge on the development of a project. If multiple strands of work are envisaged, this is the point to think about their interaction, so you have an idea about dependencies, for example, where one strand builds on the results of a second, which must therefore be finished first.

- Draft of project plan
  Once you have a good idea of the full scope of the project and the timescale, one should aim to flesh out the project plan to include a clear statement of the science goals, and the way in which the work is expected to be accomplished.

  With the full range of information that you have now assembled, this is a suitable point to use project tools to organise information and thereby understand the critical steps. This process will also help you to understand the risks associated with the project, from the scientific (is the work feasible?) through issues such as ethical clearance for work with animals or humans.

- Project plan
  The preliminary work should now have enabled you to get a clear concept of the whole project, the way in which it should be tackled and any critical hurdles.

  This information can then be used to frame a full project plan including an outline budget for the work, in preparation for the writing of a proposal. You need to think about what provision has been make for contingencies if things go wrong, a particular concern in commissioning new equipment.

  The project plan sits alongside the research proposal as a guide to the whole range of the work and its interdependencies. It can usefully include indicators of progress, such as defined *milestones*.

- Proposal development and submission to funding agency
  The particular form required for a research proposal depends on the funding agency and funding scheme. In some cases the scientific and budgetary information are separate, more commonly they appear in a single document.

  The thinking that has gone into creating the project plan helps to make the research case more convincing. You still have to 'sell' the importance and need for the scientific component in order to secure the funding.

The proposal has to conform to the rules of the funding agency, and this will place restrictions on the nature of acceptable budget items and allowable time frames. It may be necessary to put forward only part of a broad research program to stay within the bounds of what might be supported. In this case, prior planning should make the order of program components clear.

It is also important to keep track of the funding policy of the agencies to which submission is being made. Where funded amounts are regularly much less than requests, extra effort or proposals may be needed for a project to proceed in full.

- Project funded

In these days of high competition for research funding, it is likely that multiple proposal submissions are needed to secure funding. Where feedback is supplied it is critical that this be incorporated into the proposal, and the implications for the overall project plan understood.

When funding is secured, the requirements are to bring the project into action to achieve the research goals. The nature of what can be achieved will be dictated by the level of funding, and the research plan must be revisited for implementation.

- Working project plan

Now the project is about to start, the project plan needs to be revisited in the light of the prevailing circumstances, particularly the available level of funding. It may be necessary to reduce the scope of the work from what was desired, to accommodate a shorter time period, or less money. Advance planning makes such decisions easier, if not more palatable.

It is useful to retain the original project plan for reference, but now the time line and resources need to be reworked to form the basis for the way ahead.

Over the course of a project, external and internal circumstances can change, and so the plan may well need to be updated during the course of the work. Effective research management needs to recognise how the project is developing, and remain flexible to secure the best outcomes.

- Reporting, publications and presentations

Keep note of any reporting requirements and build these into the working time line. Early in the project it is worth sketching out the expected presentations at conferences etc., and which members of the research group will be responsible. Presentation material can frequently form a good base for the preparation of publications.

The final report on a funded project generally requires a summary of what has been achieved. It is worthwhile to look back on the project plan

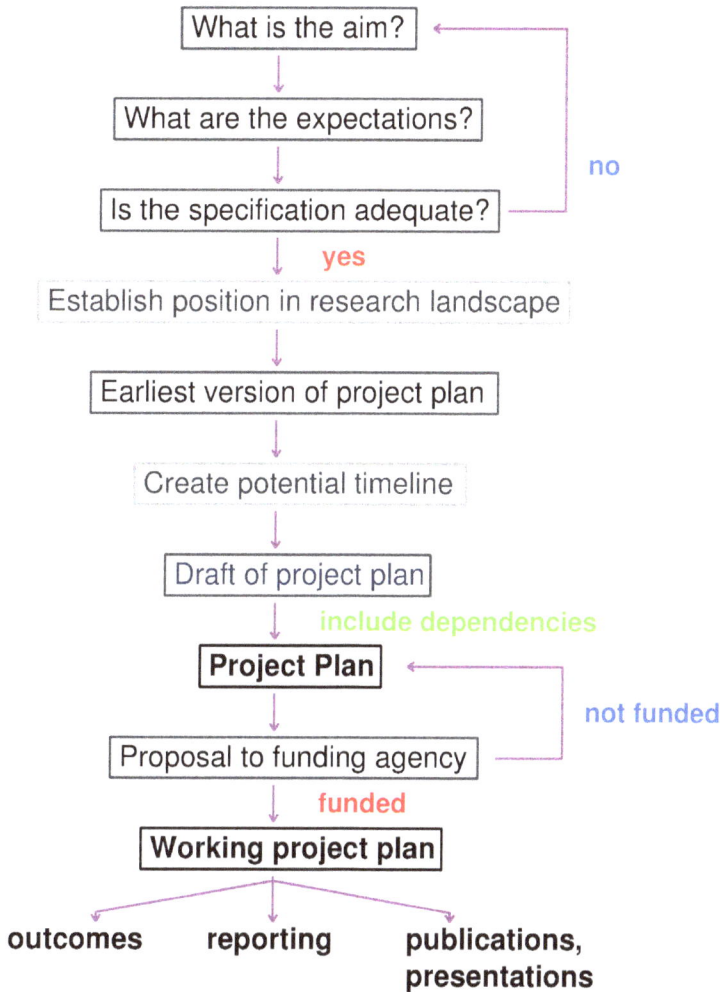

Figure 2.1: Schematic representation of the development of a research project.

at this stage and reflect on how far the forward planning matched with actual progress. This step will aid the development of the next project.

- Lessons
  Often you will find that the research has developed different aspects than expected at the beginning, and rather than answering the questions posed, you have opened up a new group of issues. This is part of the excitement of research; the outcomes can be unexpected. It is important to not force rigid adherence to a plan, except where funding is tied to specific outcomes, but to adapt to circumstances.

  The experience from one project then feeds into the creation of the next, and what has been learnt from tracking the progress of a project will help improve the next plan.

We can summarise the stages of the research project between concept and completion in the form of a flow chart, as shown in Figure 2.1.

> **Exercise 2-1:**
> Prepare an outline project plan for your current project that summarises goals and expectations together with the time frame for the major elements.

2.2 Building a research proposal

The effort put into building a project plan can be readily repaid by the way in which it eases the preparation of research proposals. In particular, a clear concept of the resources required, and the way that they will be employed, can improve the focus of the description of the science to be undertaken.

2.2.1 Formal and informal research proposals

Most research proposals are developed as a means of submission to funding agencies to secure the resources needed to implement the research. In consequence the particular format and classes of information to be required are dictated by the scheme from which funding is sought. Even within the framework of a single funding agency, requirements can differ markedly between programs, for example, with respect to presentation and budget constraints. It is therefore important to make sure that these requirements are understood and implemented in the proposal.

For moderate-size projects the normal mode of operation is to submit a single proposal, though sometimes separate documents are required for the science and budget components. Funding for larger programs will often involve a two-stage process. Initially an *expression of interest* or a relatively short proposal is required. Then selected submissions are invited to prepare a full submission of greater length. The time frames in which the full material is to be prepared can be short, and so it may be necessary to produce a first draft of the full material in case it is needed.

In some circumstances, for example, applications to some of the schemes of the European Research Council, both parts of the proposal have to be submitted with the original application. The first part is used by an assessment panel to judge whether the whole proposal is sent for review. Both parts of the proposal are seen by the external reviewers and by the panel in their final deliberations in the light of the reviews.

Even where no formal research proposal is required, it can be of value to create a short informal proposal with the aim of summarising the aims of the project and the way in which it will be tackled. This informal proposal material will be found to be valuable as a summary of initial thinking, and as a means of developing a project plan with appropriate organisation of activity. Even if external funding does not have to be sought, there will be a need to achieve the project goals. Such material provides a succinct representation of the project, which is valuable when planning future directions

2.2.2 General considerations

The aim of a research proposal is to provide a convincing case for carrying out a specific research project based on the scientific case and research capacity. The specific work component therefore has to address the following topics:

- What needs to be done?
- Why should it be done?
- Why is it important?
- What can you build on?
- How will you address the issues?
- How will you know you are right?
- What resources are needed?

The way in which these elements are represented in the research proposal will depend on the requirements of the funding scheme. Often length restrictions are imposed on the various sections of the proposal and so brief, but clear, exposition is at a premium.

In addition, information will be sought on the scientific record of the potential participants, including such items as employment history and publications. In some cases the applicant makes a separate statement about the way in which their research experience benefits the project; in others this will form part of the scientific component.

Alongside the scientific and personal elements, the proposed budget will have to be presented, with justification for the various elements. This budget needs to include both the components requested from the funding agency, and funding available from other sources. Normally the budget information is available to reviewers, but in some cases may be in a separate document that is considered only when the science review has been completed.

Increasingly, proponents are expected to point to the way in which the work proposed can provide broader societal and economic impacts. In some cases the formulation may be as simple as 'contribution to national wealth', but often the prescription is more vague, such as: 'describe how the Proposal might result in national economic, environment and/or social benefits'. Broader impacts are commonly used in proposal assessments, and so should not be neglected.

Before a proposal can be submitted there will normally be a need to obtain appropriate certification from your institution, or that through which the proposal is being submitted. The institutional deadline will normally be somewhat earlier than that imposed by the funding agency, so that the effective timescales for preparation are slightly compressed.

2.2.3 Understand your funding sources and rules

Even when submitting to a single funding agency, different schemes are likely to have differences in eligibility criteria and the specific information that is requested. It is therefore important to read the detailed instructions for the particular scheme and to follow them carefully. There may well have been changes since the last time an application has been made to a particular scheme. In particular, the details of the class of personal information can change.

Where a funding source offers multiple schemes for funding you should make sure that you are choosing the one that is most suitable for the nature of the proposed work. There tend to be more options for strategic and applied research than for basic research, since schemes may exist for collaboration with industrial or government partners. Large research initiatives may offer opportunities in a different form than the standard individual grant, but may also require closer cross-linking with other activities.

When preparing a proposal you should be aware of the *selection criteria* that will be applied to the scheme. This should guide the proportion of effort that you place on different aspects of the proposal. Generally the proposed science plays an important part in assessment, but a strong weight may also be placed on track record in the light of career circumstances.

It is also important to be aware of the detailed funding rules attached to the scheme to which you are applying: there are often restrictions on what are eligible expenses, and there may be specific limits applied, for example, to travel items.

You should also make yourself aware of the way in which the funding agency assesses proposals and makes decisions about funding. A common mode of operation is that proposals are sent for external review, and the results of the review processes are then moderated by a panel. The panel may then be involved in the funding and budget decisions, but in some cases these are made separately based on a ranking list prepared by a panel.

2.2.4 Proposals as 'sales pitch'

The proposal is the only way in which your vision and research capacity are conveyed to the review process for funding. Ultimately you are trying to sell your ideas against a background of many competing proposals. Your proposal should be as appealing as possible without resorting to overstatement or exaggerated claims. Both scientific and public benefit need to presented to best advantage.

The first thing that anyone will see is the title of the proposal, and so it has considerable importance. You should aim to make the title as clear and interesting as possible within the length constraints.

Commonly an abstract is expected for the proposal, but the required length varies considerably between different agencies. In some cases the abstract is rather short, for example, 100 words and then every word counts. The active voice should be used, not only is this shorter but it is also more direct. The major goals need to be expressed in a pithy way, so that the importance of the work is conveyed. Where a longer abstract is allowed, it is still important to present ideas as succinctly and effectively as possible with a clear emphasis on the project goals. The longer space then allows a more extensive exposition of the major steps in the project that will allow the achievement of the target.

In the main proposal one strives to make the material interesting and accessible, so clarity of expression remains very important. The structure of the material may well be dictated by the funding agency. In this case you will need to make sure that you follow the designated headings to provide suitable background to the work and to demonstrate its significance. Where the format is less prescribed it is still important to give an effective introduction with due acknowledgement of prior work, and then develop the nature of the proposed work and the way in which it will lead to the desired scientific targets.

Many funding agencies place a strong emphasis on 'testing hypotheses' as the primary role of research, and so thought needs to be given to casting the material into an appropriate form to sit within this framework.

The proposal needs to provide a clear message as to the scientific goals and the expected *outcomes*. Although many scientists can formulate clear goals, they are often less good at expressing the likely tangible results of a project. Prior project planning can help in identifying the returns from the proposed work. Do not confuse *outcomes* with *outputs*, such as publications or web materials associated with the project.

Multi-stage processes

Where the proposal process involves two distinct stages, such as an expression of interest or preliminary proposal followed by a detailed proposal, it is likely that there will be considerable differences in the nature of the material provided in the two stages. Indeed the two sets of material may be assessed in very different ways.

In order to produce a suitably crisp, short proposal for the first stage, it may be necessary to have developed a much fuller set of material. When the specifications for the second stage are already available, it is desirable to follow this as far as possible in the initial development. Some duplication in the nature of the submissions may be inevitable because of the requirements of the two stages. Nevertheless it is good practice to provide some variation in wording, against the event that both components are seen simultaneously by the same reviewer.

2.2.5 Budget preparation

The presentation of a budget is a critical part of a research proposal. The mode of presentation of the budget, particularly with respect to the level of detail, will depend on the funding agencies. Further there can be notable differences between eligibility of different classes of expenditure between different schemes from the same agency. In consequence, it is important to read and understand the funding rules attached to the specific scheme for which an application is being made. Care should be taken in providing adequate explanation for budget items within the space allotted, so that the reasons for the expenditure are well justified.

In many cases the major item of expenditure is likely to be on personnel, where costs will be linked to either to the rates of pay at the host institution or to fixed rates promulgated by the funding agency. In some cases all budgets have to be prepared at current-year prices; in others some allowance for salary increments can be included, but will need to be justified in the explanation of proposed expenditure. The arrangements pertaining to allowable travel expenditure are often quite complex, and the costs for this class of item need to be carefully assessed against the funding rules.

The treatment of infrastructure support by different funding agencies is highly variable. In some systems only direct capital expenditure on equipment is allowed; in others commissioning costs may be included. Frequently operational costs are not available in the infrastructure proposal, and so a plan has to be developed in advance as to how the infrastructure will be sustained if the proposal is successful.

Sometimes the budget element of the proposal is separate from the scientific component, so that it is not seen by reviewers of the main proposal. In this case the primary assessment will be based on the research proposed, and then budgetary decisions will be formulated separately.

Some funding agencies work on the binary principle of full or no funding; others routinely trim the allocations to successful projects. You need to understand the scenario, and the expectations and practices of the particular funding agency to which the proposal is submitted.

It is not uncommon for some funding schemes to provide less funding than requested. Nevertheless, the budget for a project should never be inflated relative to the work proposed. When the funds granted are less than the request, it will be necessary to re-prioritise the work and maybe drop some components. Such changes may require the approval of the funding agency. For agencies that regularly grant significantly less than requested, the proposal may need to build from the outset some components of the scientific endeavour that can be dropped if funds are insufficient. The full science program must justify all aspects of the work. Failure to do so will undoubtedly be noted by the reviewers and will result in negative comments and down-ranking of the proposal.

2.3 Submitting a proposal

2.3.1 Proposal presentation

Although there are many variants in the required form of proposals, there are also major features in common that are geared to the nature of the assessment process. In many cases, some material has to be entered into electronic forms, with strict character limits for sections, whilst the remainder is submitted by, for example, PDF files. Formatting instructions need to be followed or the proposal can be rejected.

The main ingredients of a proposal are:

Cover material
This includes the title, abstract and necessary institutional information

Investigators
The proponents of the proposed work have to be identified and details given of their experience and capability. There is considerable variety in the specifications of such material and the instructions from the funding scheme need to be followed closely. Interruptions to research and unconventional career paths can be identified at this stage.

> **Box 2.1:** ARC Proposal Headings for Science Component 2013
>
> - Proposal Title
> - Aims and Background
>   - ◇ Aim of proposed work, with reference to outcomes
>   - ◇ Background of research field in international context
>   - ◇ Relation to proposal
> - Research Project
>   - ◇ Significance of research
>   - ◇ How an important problem is addressed
>   - ◇ Nature of outcomes, impact and innovation
>   - ◇ Conceptual framework and methods
>   - ◇ Research plans and time line
>   - ◇ Feasibility of project - design, budget, time line
>   - ◇ National economic, environment and/or social benefits.
>   - ◇ Relation to National Research Priorities
> - Research Environment
>   - ◇ Nature of research environment in collaborating institutions and groups
>   - ◇ Relation to institutional strategic plans
>   - ◇ Communication of results
> - Role of Personnel
>   - ◇ Roles and contributions for main investigators
>   - ◇ Roles and involvement of other participants
> - References

Scientific justification

This is the main component of the proposal where a clear research goal and plan have to be expressed. It is here that the effort expended in developing a working plan for a research topic will be repaid. If the concept of the work has already been well developed with a clear expression of questions, hypotheses, and outcomes, then this can be built directly into the justification for the proposed work. This section is normally subject to fixed length limits and may also have prescribed structure. For example, Box 2.1 shows the required headings for the Australian Research Council (ARC) Discovery Scheme for 2013, together with the class of material that should appear in each section. This list gives a good indication of the topics that should be addressed in proposals, even where a more free form structure is allowed.

Budgets

The way in which budgetary information is handled varies markedly between schemes. Sometimes the budget is incorporated with the research plan, but frequently it is in a separate section. A standard budget form may be required or entries made in an electronic spreadsheet. Where optional budget templates are provided it is worthwhile using them, since it will make the information easier to digest in the review process. In all cases the budget items need to be justified carefully and explicitly. Items that may receive particular scrutiny are travel and fieldwork, equipment, consumables and maintenance costs.

Results from prior support

Proponents are often required to report on the outcomes of previously funded projects, including those for which they may not be the main investigator. In some cases, for example, the U.S. National Science Foundation (NSF) this forms part of the science component. In others it appears as a separate section of the proposal.

2.3.2 *Examples of proposal guidelines*

Some aspects of the form of proposals can be explicitly geared to the class of criteria that will be used in project assessment. Thus, in the 2013 version of the *Grant Proposal Guide* of the NSF the following requirement is included:

> Each proposal must contain a summary of the proposed project not more than one page in length. The Project Summary consists of an overview, a statement on the intellectual merit of the proposed activity, and a statement on the broader impacts of the proposed activity. The overview includes a description of the activity that would result if the proposal were funded and a statement of objectives and methods to be employed. The statement on intellectual merit should describe the potential of the proposed activity to advance knowledge. The statement on broader impacts should describe the potential of the proposed activity to benefit society and contribute to the achievement of specific, desired societal outcomes. The Project Summary should be written in the third person, informative to other persons working in the same or related fields, and, insofar as possible, understandable to a scientifically or technically literate lay reader. It should not be an abstract of the proposal.

The guidelines for the *Starting Grants* of the European Research Council provide a useful summary of expectations with respect to the expression of the objectives of the project and necessary background, as well as the project plan:

> *a. State of the art and objectives:*
> Specify clearly the objectives of the proposal, in the context of the state of the art in the field. When describing the envisaged research it should be indicated how and why the proposed work is important for the field, and what impact it will have if successful, such

as how it may open up new horizons or opportunities for science, technology or scholarship. Specify any particularly challenging or unconventional aspects of the proposal, including multi- or inter-disciplinary aspects.

b. Methodology:

Describe the proposed methodology in detail including, as appropriate, key intermediate goals. Explain and justify the methodology in relation to the state of the art, including any particularly novel or unconventional aspects. Highlight any intermediate stages where results may require adjustments to the project planning.

Note that it is expected that the project plan be dynamic, and therefore able to adjust to the nature of earlier results. It is not a weakness to anticipate critical points in the work schema, and to have alternative modes of attack depending on outcomes.

2.3.3 Before submission

Proposals need to be written in a simple, direct and concise style with a clear logical flow. Try to avoid over-dependence on highly technical language, though occasional use of jargon is almost inevitable. The use of subheadings can be helpful to improve the organisation of the material. A careful check should be also be made for instances of poor expression and grammar, and spelling errors. In short, you need to read the proposal as well as write it. Setting the material aside for a few days before reading afresh can help.

Figures prepared for publication may well be over-complex for a proposal and need too much explanation. Figures should concentrate on material that is important for understanding the new work proposed, rather than older work already accomplished. Compressing a figure into a small space can compromise the quality in the version seen by a reviewer. For a complex project a diagram indicating the interrelations of the different components can be helpful.

You need to start early enough so that there is sufficient time to carry out the proposal writing and necessary revisions, as well as seeking feedback from your peers. It can be particularly helpful to get a reaction from someone who is not in your field.

2.4 Proposal review

Once a proposal is submitted there is a natural inclination to relax and concentrate on research once again. But, in the meantime, your proposal will be starting on its track through the review process. It is helpful to understand the nature of the reviewing system, since this can assist in

---

Box 2.2: ARC Discovery selection criteria 2013

a. **Investigator(s)** (40%)

   ◇ Research opportunity and performance evidence
   ◇ Time and capacity to undertake the proposed research.

b. **Project Quality and Innovation** (25%)

   ◇ Does the research address a significant problem?
   ◇ Is the conceptual/theoretical framework innovative and original?
   ◇ Will the aims, concepts, methods and results advance knowledge?

c. **Feasibility and Benefit** (20%)

   ◇ Do the projects design, participants and requested budget create confidence in the timely and successful completion of the Project?
   ◇ Will the completed project produce innovative economic, environmental, social and/or cultural benefit to the Australian and international community?
   ◇ Will the proposed research be value for money?

d. **Research Environment** (15%)

   ◇ Is there an existing, or developing, supportive and high quality research environment for this project?

---

improving the design of projects. Experience gained in reviewing enables one to see a proposal in a different context. Even informal review of other people's projects can open your eyes to different styles of presentation and expression.

The first step in the formal process of review at a funding agency is the selection of projects that will be reviewed and the nomination of suitable reviewers. Such a process may involve a program manager or an expert panel.

The next task for the funding agency is to secure a sufficient number of external reviews. Normally at least two reviewers will be involved, but sometimes up to ten may be sought, particularly for very large projects. With the growth of scientific endeavours many calls are made on scientists' time and, in consequence, the rate of acceptance by nominated reviewers can be low. Certainly at the earlier stages of one's career one should accept such an invitation to review a proposal, if at all possible, because it represents an effective learning experience that may well help you in your own proposal writing. Further, you depend on the willingness of others to review your proposals, and so some reciprocity is required.

Conflict of interest rules generally preclude review of proposals involving recent collaborators, former students or advisors, and also from your own current or recent former institution or an institution with which

> **Box 2.3:** NSF Proposal Review Criteria 2013
>
> When evaluating NSF proposals, reviewers will be asked to consider what the proposers want to do, why they want to do it, how they plan to do it, how they will know if they succeed, and what benefits could accrue if the project is successful. These issues apply both to the technical aspects of the proposal and the way in which the project may make broader contributions. To that end, reviewers will be asked to evaluate all proposals against two criteria:
>
> - **Intellectual Merit:** The Intellectual Merit criterion encompasses the potential to advance knowledge; and
> - **Broader Impacts:** The Broader Impacts criterion encompasses the potential to benefit society and contribute to the achievement of specific, desired societal outcomes.
>
> The following elements should be considered in the review for both criteria:
>
> 1. What is the potential for the proposed activity to:
>   a. Advance knowledge and understanding within its own field or across different fields (Intellectual Merit); and
>   b. Benefit society or advance desired societal outcomes (Broader Impacts)?
> 2. To what extent do the proposed activities suggest and explore creative, original, or potentially transformative concepts?
> 3. Is the plan for carrying out the proposed activities well-reasoned, well-organized, and based on a sound rationale? Does the plan incorporate a mechanism to assess success?
> 4. How well qualified is the individual, team, or organization to conduct the proposed activities?
> 5. Are there adequate resources available to the PI (either at the home organization or through collaborations) to carry out the proposed activities?

you have a regular working relation. The arrangements differ between schemes, but normally you will need to raise any potential conflict with the funding agency who may then determine whether the review can proceed.

Before starting the review it is important that the reviewer understand the general requirements of the funding scheme, and the nature of the assessment criteria. In some cases the selection criteria are presented explicitly — an example is provided for the Australian Research Council Discovery Scheme in Box 2.2 — and tied to the nature of the questions posed on the review form. Sometimes the information is effectively contained in the entries that are requested in an electronic review form. In other situations the merit criteria are clearly presented but the way in which they will be employed is not available to a reviewer. For example the U.S. National Science Foundation provides a detailed set of general instructions to reviewers (Box 2.3), but the relative importance to be attached to the different components may vary with funding scheme.

Whatever the criteria being employed in the assessment it is important that the review process is fair, and judges a proposal on what has been written. Where critical comments are made they should be substantiated with appropriate information, rather than a bald statement such as 'this cannot possibly work'. A review should look for true innovation and give it due praise. In many cases both written comments and numerical or other rankings are required. These two different aspects of the assessment need to tally if they are to be used effectively.

Although it is common for reviewers' comments to be passed back to the proponents in some form, this should not inhibit a reviewer in making justified criticisms nor making constructive suggestions for improvement. Exaggerated language, either positive or negative, is not helpful for those who will use the review.

Some funding schemes invite commentary on the suitability of the budget, but others separate assessment of scientific merit and budgetary considerations. When assessing budgets you should be aware that the circumstances prevailing for the proponents may not correspond to those with which you are familiar. Costs are assessed in different ways in different countries, and often reviewers are not provided with the full instructions as to eligibility of costs.

Some funding agencies allow a response ('rebuttal') to the reviewers comments. Where available this should be viewed as a chance to seize on the positive aspects of the reviews, and also to blunt criticism. If a review is wrong then this needs to be pointed out politely, and the correct explanation given. The reviews and the responses will be commonly be seen at the final assessment meeting by panellists who will also have had an opportunity to see the proposals. This review panel will normally determine ranking lists, and may also have a role in determining funding.

Exercise 2-2:
Prepare a personal profile in the style required by your preferred funding agency. In some cases this may constitute a simple curriculum vitae, but others will have a more complex structure; e.g., including your best 10 papers with reasons for the choice.

Exercise 2-3:
Obtain a copy of a submitted proposal in a field different to your own and prepare a review based on the appropriate criteria for your preferred funding scheme, with both numerical scores and written comments (use a 1-5 scale, with 5 best).
This exercise is most effective if you can exchange and critique proposals with colleagues from another field.

2.5 The project proceeds

Congratulations, your project has been funded. Now you have to turn your concepts into reality. The first stage is the interface with your institution to get necessary agreements implemented, and administrative arrangements set in train. While this is happening it is a good idea to develop your project plan in full, with well-developed time lines and structures so that you can manage the progress of the project. In the next chapter we discuss the various considerations that need to be taken into account, and the project management tools that can aid the task.

Often you will want to recruit staff and this will require a different interface with the Human Resources structures so that advertising and appointment processes confirm to institutional rules. Students will commonly be involved with the project and this again invokes a different set of administrative requirements.

The main goal remains to pursue the scientific enterprise, and so it is good to get the work underway as soon as is practicable. You need to bear in mind needs for any intermediate reporting, which is often annual for major funding agencies, but more frequent with industrial sponsorship. Such milestones can be built into your project plan, so that they do not come as an unpleasant surprise that leads to interruption of the main effort.

Increasingly funding sources require explicit data-management plans with accessible archiving of data, so the protocols should be established at the beginning of the project to ensure that the appropriate *metadata* is registered. This is as much for your benefit as anyone else. If you need to go back on your tracks, the maintenance of good records by the entire project team will allow a swifter response to the changing direction of the work.

Hopefully, the project proceeds smoothly, but surprise is a constant factor in successful research. You therefore need to be prepared to exploit new insights and results with modifications to your plans. It is worth making a quick check before a change of direction to ensure that obligations associated with the funding will still be met, and also that there are no budgetary hurdles.

2.6 Project completion

As the funding period draws to a close there are a number of issues that need to be considered. Generally, submissions for additional funding for a cognate or different type of project will have been made by this time. The circumstances will then vary depending on whether new funding is available, or the current funding terminates and *ad hoc* measures have to be taken to maintain research.

Whatever the future state, you still have obligations to fulfil on the current project. Commonly a final report has to be rendered to the funding agency detailing the work achieved, the research outputs such as publications, and data management. This is the stage at which you will be trying to exploit the work achieved through conference presentations and publications with a broader overview. Good record keeping through the project, and thorough *metadata* attached to experimental results or simulation procedures, will aid the process of producing these major outputs.

The other step at the end of a project is the finalisation of financial accounts, so that expenditure is fully justified. Since financial reporting may well take a different path through an institution than the handling of research reports, you will have to ensure that necessary communication occurs. Many people's experience indicates that this is not something you can take for granted.

Chapter 3
How to Plan and Manage a Project

In the previous chapter we have discussed the development of a research concept, and the way in which this evolves into a specific plan that can form the basis of a research proposal. Now we start to discuss the ways in which research planning can be systematised, and used to manage the activities for one or many projects.

We start by considering simple projects where the process is linear, and then move to more complex situations involving interdependencies between components of activity. The simple case can be dealt with by a structured task list in an *action plan*. Once multiple strands of work are involved it becomes important to specify suitable intermediate goals, and to bring all the information on the project together to develop a clear time-line. A powerful tool in this context is the *Gantt chart*, which summarises the resources available and the way that they can be deployed to achieve the ultimate goal. Effective planning requires the recognition of the way in which the different components of the work interact. The result of this interaction is that later activities can be strongly affected by earlier components without this being immediately obvious. The use of *critical path analysis* is designed to bring out the impact of such complications, and to aid the removal of obstacles. The necessary information can be extracted from the Gantt chart, but needs to be organised to provide the maximum information.

Since even the best laid plans are subject to unexpected influences, it is important to be able to assess the real progress in a project, so that the project plan can be updated if needed. Assessing progress requires some pre-assigned markers against which the actual situation can be compared. Commonly two classes of markers are employed: (i) *milestones* that represent the achievements of the project that have to be accomplished by specific dates, and (ii) *key performance indicators* that set a range of criteria for which quantitative measures can be developed. Milestones are generally closely linked to the intermediate goals of a project and are focused on outcomes. Performance indicators can be arranged to take more account of outputs such as publications

A component of planning that is receiving increasing attention in research contexts is the assessment of *risk*, which may be technical, financial or even involve reputation.

3.1 Project planning

In Chapter 2 we discussed the way in which a project plan can evolve as a concept is brought towards realisation and funding. The earliest stages of such a plan aim to gain an overview of the whole project so that the necessary resources can be brought to bear to achieve a successful outcome. In particular it is important to identify any dependencies between components of the proposed activities, so that not only are they carried out in the right order, but also bottlenecks are avoided. Normally, up to the point of proposal submission, one will be concentrating on the main features rather than the details. However, for very large programs with multiple components the program plan required for successful proposal submission will need to be developed in some detail.

Once a project is funded then the concepts have to be turned into a working plan, which will need to allow for change to ensure that the goals are met. The working plan needs to include more detail than is needed in a proposal, so that all the strands of activity can be brought together. It is important to recognise that the features recognised in the proposal may not represent the whole project, since there may be dependencies on, for example, external facilities whose scheduling is not under the control of the project. Further, the goals may have to be scaled back in the light of available funding.

The working plan should build on the information developed for the proposal, particularly for such issues as the way that different components of the project interact, and where bottlenecks may occur. Critical path analysis can help to recognise such features, and suggest ways in which they may be overcome.

The stages of planning need to encompass both an appropriate time line and effective use of resources:

Time lines. Based on an understanding of the whole project it is important to identify both intermediate and final deadlines. The final deadline may be externally imposed, but commonly the intermediate steps will be under your control, and so some flexibility should exist in assigning deadlines. In the construction of a time line you need to think about whether there are aspects of the project where delays could occur. If so you need to build in contingency plans, so that the whole project is not held up waiting for a particular result. Commonly the main control on how fast work can progress comes from the availability of people, and how their efforts can be assigned across the necessary work.

Costing issues. It is always salutary to consider how much the project outcome is worth. This question is often difficult to answer, but it can help to put a disappointing funding result into perspective. You will know how much funding you would like to have and, if faced with lesser support from a primary funding source, you need to consider your courses of action. It may be possible to scale back the enterprise with reduced goals. Otherwise you need to seek other sources of support.

Resources. A successful research project makes good use of available resources. Issues that need to be considered are the availability of suitable infrastructure. If equipment is not accessible, do you need to build in acquiring and commissioning new equipment as part of the requirements? If so, you may need to develop a bid for infrastructure support in parallel with, or before, the main research proposal. Does the necessary expertise for experiments or analysis exist? Do you have to allow time for extended sample preparation or, in more theoretical work, the development of necessary software? Are there adequate personnel resources or will new staff have to be recruited? It takes time to advertise, recruit and start someone working on a project and this must be taken into consideration. If the final deadline is somewhat flexible, it may be best to delay the start until everything is in place. Otherwise you may be forced to start anyway, and pick up speed when the new staff arrive — with a consequent update of the working plan.

3.2 Simple projects

Many projects appear to have a simple structure in which each element leads logically to the next. Nevertheless, it can be easy to overlook an apparently minor component that has a strong influence on the outcome.

An *action plan* represents a list of all the tasks that need to be accomplished to achieve an objective in the appropriate sequence. Such a simple list provides a framework to guide thinking about how to complete the project effectively. Once all the steps are laid out it is easier to recognise the relative importance of each component and where, for example, extra help may be needed. Alternatively, some tasks may be downgraded in importance and ultimately carried out in parallel with the main activity.

To draw up such an action plan, list the tasks that you need to complete to deliver your project or objective, in the order they need to be completed. This requires careful consideration of the steps required to bring the project to completion. Once all the likely components are assembled see if any can be pruned, or alternatively require more than the expected availability of resources.

In parallel with the task list, consider the time line, so that simple milestones get attached to the tasks in the list. Further, take account of the availability of such items as equipment and people in relation to the time-line. At this point you may recognise that even a well-formulated project may not be simple in practice. As soon as you need to schedule people's time, or meet tight deadlines as part of your project, then it is worth using more formal project management techniques.

3.3 Project management tools

It is often feasible to get an good idea of the important aspects of a project without explicit use of project management tools. However, the use of such tools forces a very close inspection of what is required, and what is available. Indeed the work of building the project plan can reveal unexpected issues.

The process of organising and managing a project can be greatly aided by making use of standard tools for which there is well-developed methodology. Such use does not necessarily require significant expense because there is high-quality freeware available. The commercial software versions are geared to larger programs with a number of sub-components, and provide more sophisticated graphic output.

Management tools provide tracking of progress, based on assignment of tasks, personnel and resources. In all cases, effective use of the tools requires a good understanding of the specifics of the project. Indeed, the analysis required to set up the tools frequently provides significant insight into the project itself. As you will see, it is important to understand the way in which different parts of the project interact. Such knowledge of dependencies allows the identification of the *critical path*, i.e., the set of steps that determine the rate of the whole project. If possible, effort can be directed to these critical steps to smooth the passage of the total effort.

The tools we are about to introduce enable the construction of a suitable project plan, and allow progress to be judged. However, it is important to remember this is just a *plan*, which needs to be adjusted as circumstances change. The plan should be updated from time to time as the project develops, but it is desirable to retain earlier versions for reference

3.3.1 Breakdown of tasks and resources

The first step in developing a good project plan is to identify the tasks to be accomplished to achieve the desired research outcomes, and the available resources. This is the point at which you start to estimate the time required to complete each task.

A way to begin is to write a set of headings for each of the desired outcomes, and under these list the components of the work that need to be done, subdividing if necessary where there are multiple components

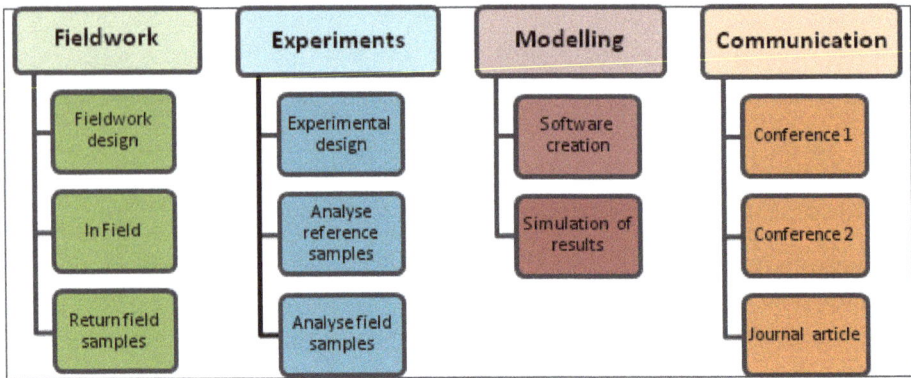

Figure 3.1: Work framework for a multi-strand research project.

for a particular component. This can be achieved with paper and pencil, or with post-it notes on a whiteboard, which works well when a group of people are involved. You may find it easier to start with a simple list and then organise under headings. The final product should then be captured in a spreadsheet or similar style. At this stage the order in which the tasks are to be performed is not important – that will come in the next step. It would be rare for a research project to need to go beyond a second level of sub-tasks. Large programs may need many more since the first level may well be aggregates of projects.

An example for a moderate-size project, with multiple strands of work including field data collection, experimental work and simulation, is shown in Figure 3.1. At this stage the major tasks that need to be accomplished for each strand of the work are identified. At the next stage you will start to include the inter-relations between the strands.

Once this framework for a project is assembled you need to check that all the elements of the necessary work are included. If not, add to the lists until all aspects are covered. Also start to think about any necessary cross-linkages between the different elements. This may sound complex, but as the framework is sketched, many of the relevant issues tend to come to mind.

In parallel, you need to build a list of resources that will be required to enable the tasks to be accomplished. You need to think about the availability of people and equipment, the time frame needed for such aspects as data collection, laboratory analysis or fieldwork, and the way in which the various tasks may interact.

The identification of the various tasks helps you to assess how you can deploy the resources over the available time span. Such planning in the pre-proposal stage aids with developing an appropriate sense of the time and budget needed for a successful project.

Once the project is funded the planning can become more detailed. We will see shortly how this information on the necessary tasks and their time relations can be presented in a readily assimilable form using Gantt charts that express both the time lines and the interdependencies of the project components.

3.3.2 Project schedule development

With the aid of the work framework you now need to consider the sequence of activities and the way in which the different elements interact. You also need to start to specify the duration of each activity so that the total nature of the project becomes clear. Estimates of project duration will depend on how many people are involved and what proportion of their time they can devote to the work, as well as availability of necessary equipment. It is easy to underestimate the time required for a particular task. You are therefore well advised at first to seek advice from those with more experience.

You also need to think about the temporal relations between tasks. Some can only start when other tasks have started, or when some other activity has finished. Completion of some aspect of the work can depend on other tasks finishing. A few components may exist separately and can then be slotted in where appropriate.

It is useful to rough out the sequence of activities in approximate chronological order, including the dependencies between tasks. This done in Figure 3.2 for the project we used in Figure 3.1. We see that it is important to get the fieldwork component completed early in the project so that the maximum time is available to analyse the samples returned from the field.

Once the time relations between the elements of the project become clear, the issue of the duration of each of the elements needs to be defined. This enables the time sequence to be refined so that the most effective use is

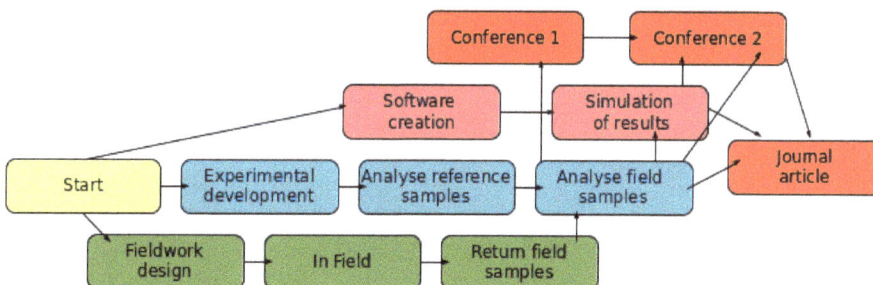

Figure 3.2: Work framework including time relations for the multi-strand project shown in Figure 3.1.

made of resources. It is at this stage that you can often begin to recognise the *critical path* for the project representing the time-limiting steps.

3.3.3 Gantt charts and critical paths

Gantt charts provide an oversight of a project in the form of a calendar view, that allows the sequence and duration of the different tasks to be appreciated. They are a good tool to work out and display the minimum time required for a project, and which tasks need to be completed before others can start. A Gantt chart also helps with recognising the critical path – the sequence of tasks that have to be completed on time if the whole project is to meet the final deadline.

The Gantt chart provides an easy-to-read representation of the project at any point and can be used to monitor progress against the plan. In particular the chart enables one to see how project milestones link to the project components. However, if there is a need for change, it can be difficult to assess from a static chart the impact of what happens in one area on the rest of the project. Most Gantt charts are now produced using computer software and so it is possible to adjust the parameters and see what happens. It is, of course, desirable to work on a copy rather than overwrite the main description of a project!

Although simple Gantt charts can be constructed using a spreadsheet, considerable flexibility can be achieved by using specific software. A number of different styles of program are available, with varying levels of sophistication. For typical scientific projects the full power of major commercial software is rarely required. The example used in Figures 3.3–3.6 has been constructed using the freeware GanttProject, which is a Java-based program that can be employed on multiple operating system platforms.

The construction of a Gantt chart requires the specification of the beginning and end points of each component of the activity. In full use, the resources to be attached to each element of the work will also be included. Milestones can be superimposed on the basic information.

For many scientific projects the personnel available will remain constant over the course of a project, with perhaps some delay at the beginning where recruitment is necessary. However, access to equipment can represent a component that requires careful scheduling and management.

We continue with the example of the multi-strand project introduced in Figures 3.1–3.2, and now prepare the associated Gantt chart (Figure 3.3) based on a two-year duration. I have used a similar colour scheme to that in Figure 3.2, so that the different strands of the work are clearly visible.

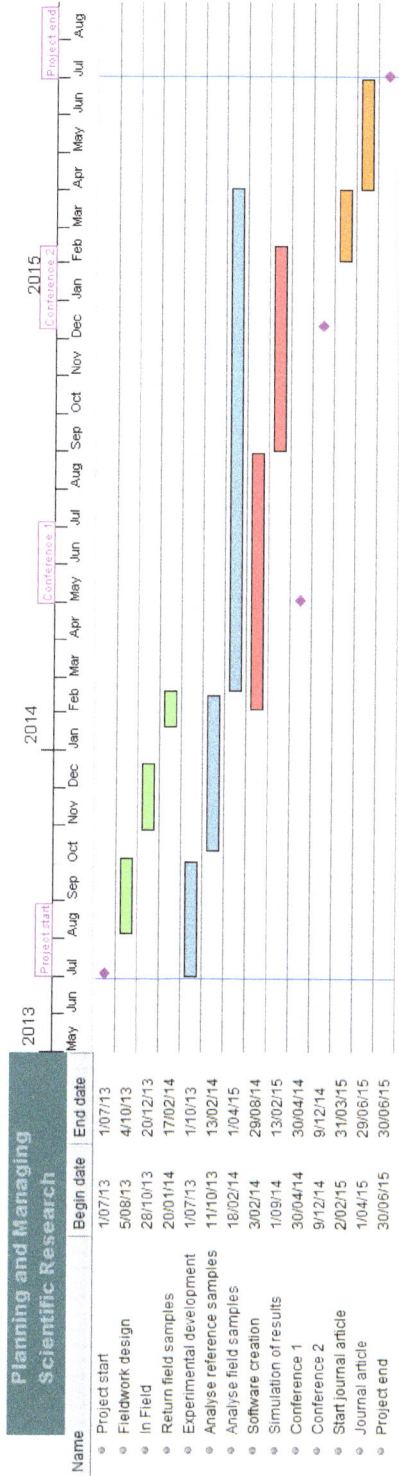

Figure 3.3: Gantt chart for the multi-strand project illustrated in Figures 3.1–3.2.

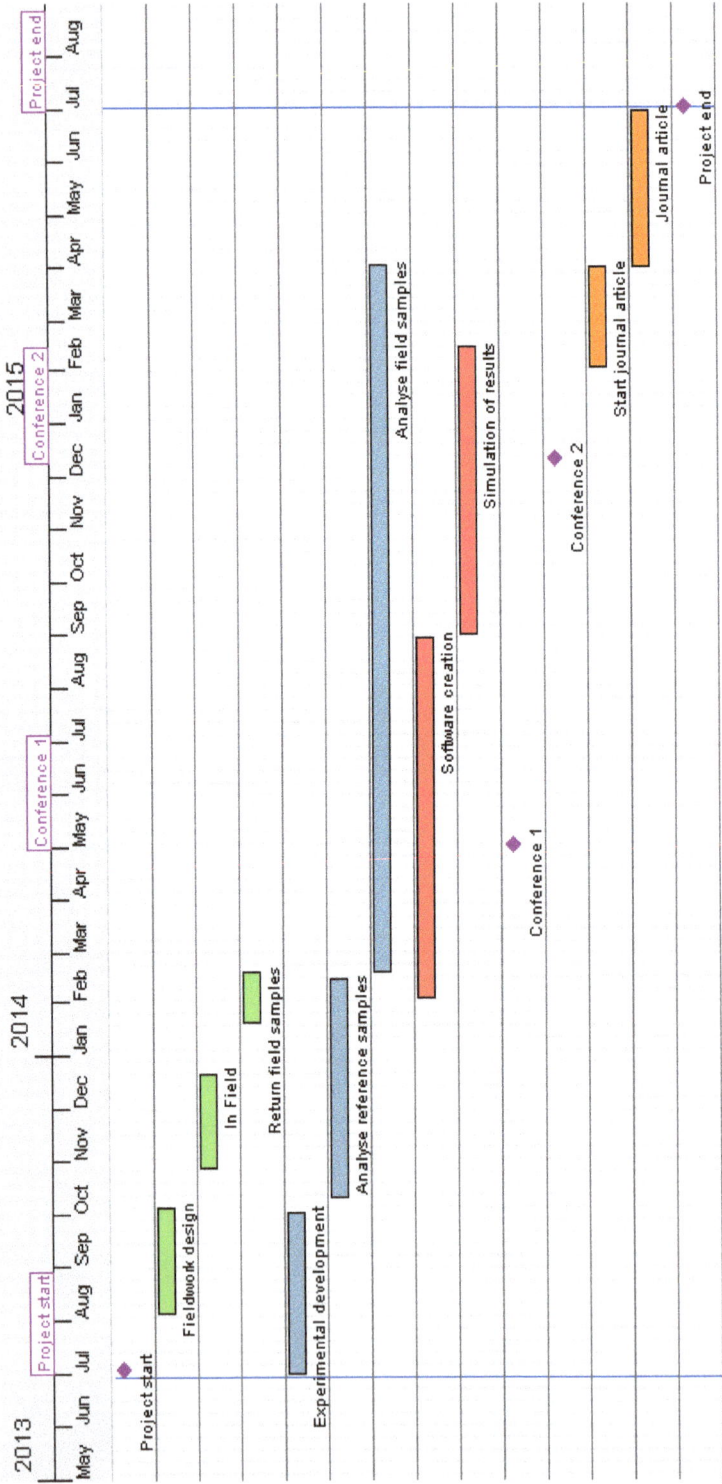

Figure 3.4: More detailed view of the project elements and timeline of the Gantt chart for the multi-strand project illustrated in Figures 3.1–3.2.

The time progression of the project and the interdependencies are developed in Figure 3.2, for example, field samples cannot be analysed until they get back to the laboratory. At the stage of development in Figure 3.2 the duration of the elements was not specified. Now we have to start putting in start and end dates for the different components of the project, so that the entire work program can fit into the specified duration. At this stage, you need to take into account not only constraints arising from the project itself, but also external influences, including staff leave. In Figure 3.3 I have included the strong effect of the southern hemisphere Christmas. The fieldwork has to be completed in time for Christmas and shipping of samples is likely to be delayed until services start to resume in mid-January. Punctuating the project are the conference presentations, which I have specified as milestones, along with the formal end of the project.

Since Figure 3.3 includes a large amount of information in a small format, I have extracted just the time block information in Figure 3.4, with now the time blocks annotated with nature of the work. We now have the time-line for the project adequately specified, but have not yet included the way in which the elements of the work interact, even though this can be sensed by the way that the strands of the work are organised.

In the project description we can now specify the predecessors of a project element, including how they are related, for example, the start of one element depends on the completion of another. Alternatively you may wish to have two elements finishing simultaneously. Time delays can be incorporated to allow for such effect as the Christmas break. Also the strength of the relationship between project elements can be specified, where this is strong then the second project is forced to move to accommodate the first. The software endeavours to adjust the specified start and end dates so that the work can be accomplished in the desired order. This may mean that the end date of the last element pushes out beyond the completion date. In which case, you need to be able to work out where adjustments can be made.

This brings us to the concept of the *critical path* which represents the combination of tasks that must be completed on time for the whole project to be completed on time. Once this critical path has been identified, then one can see which tasks can be delayed if resources need to be reallocated to catch up on missed or overrunning tasks.

In Figure 3.5 I have introduced the interrelations between project elements for our illustrative multi-strand project, using the relations indicated in Figure 3.2. Some adjustment of delays between elements was needed to avoid awkward time shifts from the plan specified in Figure 3.4. The presentations at conferences have been included as simple milestones,

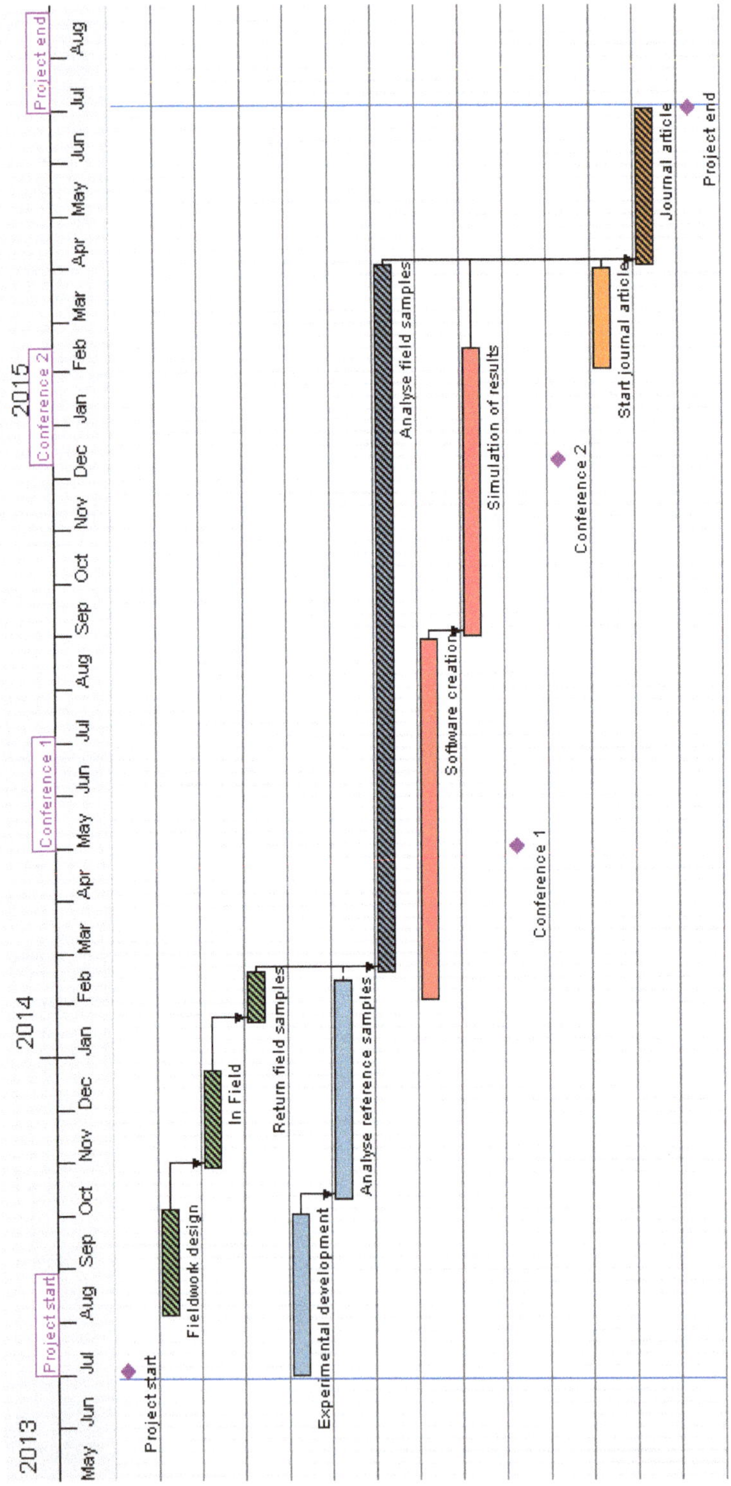

Figure 3.5: Gantt chart with dependencies and critical path for the multi-strand project. The critical path is indicated by the shaded project elements and heavier line.

and thus distinct time markers for progress, though it would be possible to include links from the project elements.

The critical path in Figure 3.5 is shown by a heavier line between project elements and the critical elements are themselves shaded. Although the writing of a journal article was intended to be the culmination of the project, the planning made it clear that delaying until all analyses and simulations were complete would lead to a rather rushed scenario. Consequently the preparation of the journal article was split into two parts; the first building up the framework of the article, and the second which depended explicitly on the analyses. The critical path, as is common, is controlled at first by a single strand of the work (in this case the Fieldwork strand leading to return of samples) and then moves over to the experimental work, before ending on the write-up of the project in the journal article. Notice that the Modelling strand has, at this planning stage, no influence on the critical path. This is because it has been assumed that all aspects will be accomplished within the specified time frame. If, however, the software development takes longer than expected then the displacement of time lines could lead to a change in the critical components for the project.

PERT (Program Evaluation and Review Technique) charts provide an alternative way to represent the dependencies and interactions of more complex projects. They include the information on start and end dates for each component, and the way that these feed into each other. Commonly such PERT charts can be produced using the same software as the Gantt charts. The example in Figure 3.6 for our multi-strand project was produced using GanttProject, making use of the interactive feature that allows improved layout over that produced automatically.

It may seem to be too much effort to go to the trouble of creating such a framework for a project, rather than proceeding and adjusting as necessary in an *ad hoc* way. Nevertheless, there is considerable value in trying to set up a detailed project plan. Frequently the thinking required to identify the elements of a project and their interrelations reveals potential stumbling blocks or bottlenecks that can be addressed before they become a problem. The time line can form a useful discussion point for the members of a project, and feedback can often help to improve the project outcomes.

The software packages that enable the construction of Gantt charts and the critical path also allow these aids to be used dynamically to include the progress of the work through the project elements. It is best to save the original version for reference. Then a working version can be used to monitor progress and to make necessary adjustments when project elements fall behind, or indeed get ahead of the expected schedule. Whilst valuable, this approach is probably more appropriate to a large program with many sub-components, rather than a typical research project.

Project start	
Start: 1/07/13	
End: 1/07/13	
Duration: 0	

Fieldwork design	
Start: 5/08/13	
End: 5/10/13	
Duration: 45	

In Field	
Start: 29/10/13	
End: 24/12/13	
Duration: 40	

Return field sam...	
Start: 21/01/14	
End: 19/02/14	
Duration: 21	

Experimental dev...	
Start: 1/07/13	
End: 2/10/13	
Duration: 67	

Analyse referenc...	
Start: 11/10/13	
End: 14/02/14	
Duration: 90	

Analyse field sam...	
Start: 19/02/14	
End: 3/04/15	
Duration: 292	

Software creation	
Start: 3/02/14	
End: 30/08/14	
Duration: 150	

Simulation of res...	
Start: 1/09/14	
End: 14/02/15	
Duration: 120	

Start journal artic...	
Start: 2/02/15	
End: 1/04/15	
Duration: 42	

Journal article	
Start: 3/04/15	
End: 2/07/15	
Duration: 64	

Project end	
Start: 30/06/15	
End: 30/06/15	
Duration: 0	

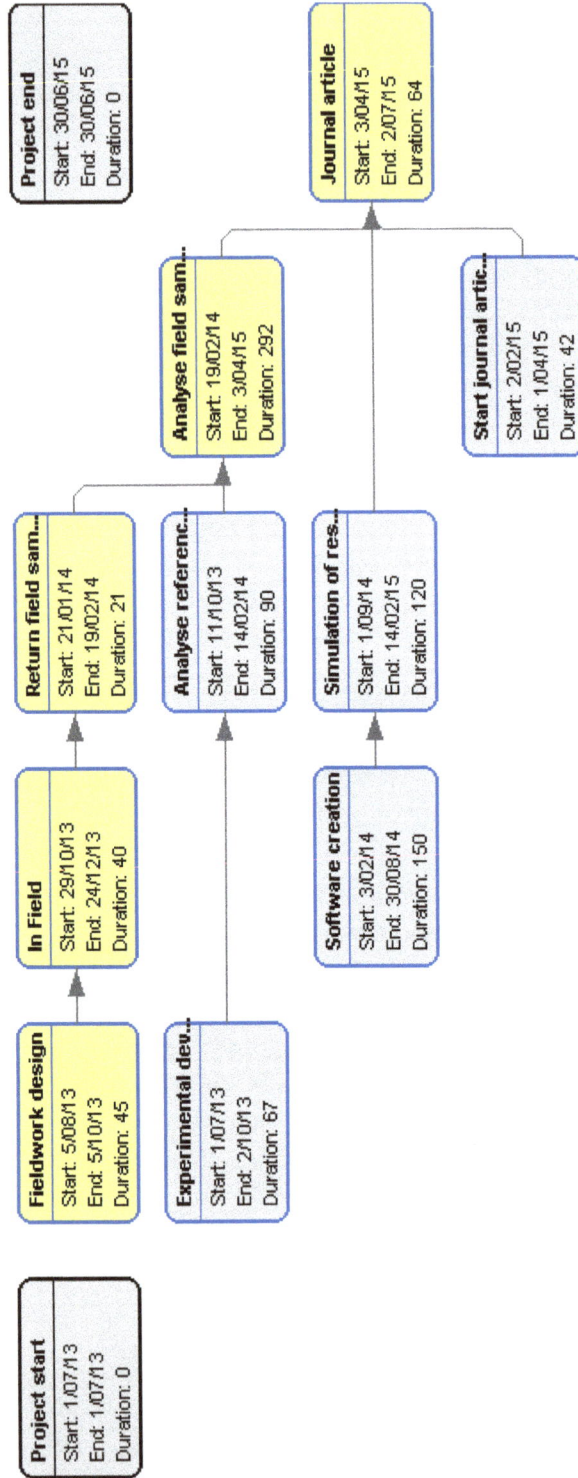

Figure 3.6: PERT Chart for the multi-strand project. The critical path is here marked by the yellow project symbols.

For moderate size projects, I have generally found it adequate to have a look at progress every couple of months or so and then, if need be, prepare a revised summary of the project development with an update of the Gantt chart and reconsideration of the critical path.

3.4 Tracking real progress

Even if project management tools are not being employed it is important to keep track of the progress of a project, so that problems can be identified before they get out of hand. In all cases it is important to maintain good records of activity, and documentation of both successes and problems. In the beginning this may seem excessive, but when one has to juggle multiple time commitments, even brief notes can be of considerable value.

Tracking the progress of a project is helped by well-conceived milestones and targets. One is then able to judge whether important stages of the project have been achieved on time or, if not, what action is required to catch up. In a multi-year project you may need to revise the milestones and targets for later years to reflect changing circumstances as the project develops. This is certainly feasible if the project is under your full control, but may not be possible without re-negotiation if milestones have been agreed with an external funding agency.

In assessing progress it is important to be realistic about project successes and delays. The progress of research is uneven and, by its nature, carries with it the risk that a concept may not be feasible. It may be necessary to retreat and abandon an unproductive approach, before trying a new way around a problem. In consequence, contingencies should be built into the plan, particularly for stages where the development of the plan suggests that progress is not guaranteed.

In a multifaceted project not all components can proceed at the same pace. It is therefore important to crosscheck progress against earlier estimates. You need to recognise potential bottlenecks, for example, via critical path analysis, and to concentrate on the performance with respect to such issues. As the project develops the nature of the critical points will change, and so management has to be dynamic rather than static.

Financial progress

Alongside the progress of the scientific aspects of the projects, there is a need to ensure that the financial state is sound and that the 'burn rate' for funds is not too high to be sustainable.

Many financial systems are geared to institutional needs and legal requirements. As a result they may not be well suited for tracking the detailed progress of a project. Personnel costs will normally be a major component and these are the most predictable. Other items may take time to enter into the financial system. Financial reports tend to be retrospective so that the current state is difficult to determine. The treatment of future commitments is tied to the specific accounting system being used. Where this is cash based, items only appear when an invoice is raised or paid. Systems with an accrual component are more complex. It is worthwhile making some effort to understand how the accounts are handled so that you do not reach the embarrassing, and difficult, situation of running out of funds on a project.

3.5 Measures of performance

There is a well developed set of criteria in project management for measuring performance against agreed goals. The main components employed in assessment of research programs or research infrastructure programs are the specification of *milestones* and *key performance indicators*. Such material may be required for major programs as part of the contractual obligations for funding, and indeed continuation of funding may be conditional on performance.

When first encountered such performance measures seem to be alien to the research environment. Yet, thinking explicitly in terms of the major steps in a project and the desired outcomes can be of considerable value when framing a research proposal, and in tracking progress.

3.5.1 Milestones

Milestones indicate the key dates to be met during the execution of the project. They should mark specific phases of a project. It is best to choose the timing of the achievement of clear tangible outcomes, rather than the preparation of outputs, such as publications, even though these may be important. Reporting deadlines also form important milestones.

In many cases specification of both start and finish dates is required, which are then to be compared with the actual situation. Even where a milestone lies well into the future, intermediate reporting is possible through the percentages of accomplishment of the task.

When the time interval for a component of the work lasts longer than a financial year it may be necessary to introduce sub-milestones, for example, once again based on achieving a fraction of the desired results. For large programs, milestones are commonly agreed with the funding sponsor and take on greater significance.

When well chosen, milestones can be an effective tool to maintain momentum in a project. Too often they are regarded as a nuisance or 'millstone'. Yet, in the early stages of planning you will have identified the major steps and the milestones are just the link to effective scheduling.

3.5.2 Key performance indicators

Performance indicators provide another way of assessing the progress of a project. They are designed to measure the extent to which the goals are achieved, and so tend to look outward from the project, whereas milestones are more inwardly focused. In consequence, performance indicators can take a broader perspective than used for milestones, and also be more oriented to outputs. A simple example of a performance indicator is what percentage of milestones are achieved on time (and/or within budget)? For an infrastructure project, suitable indicators can be the level and nature of usage as well as the tangible products such as publications

The author was heavily involved in the planning and execution of a major infrastructure program in Australia in the Earth Sciences. The AuScope infrastructure program formed part of the National Cooperative Research Infrastructure Strategy (NCRIS). The Department of Education, Science and Technology of the Australian Government put forward a range of performance indicators that could be adapted to the specific circumstances of the different capabilities being funded under the scheme. These indicators were to be used in monitoring annual performance across the various strands of the infrastructure development and provision to users. The list is given in Box 3.1 and covers a broad range of aspects of the program, not just the building of the equipment but also its quality. A major emphasis of NCRIS was on fostering collaboration. As a result collaboration plays a significant role in the performance criteria, but notice how it has been coupled with the need to achieve world-class research.

3.6 Reporting requirements

For most research projects the primary outputs are publications in research journals, and these constitute the main way in which the work is reported. However, many funding agencies require a final report at the end of the project, and for larger programs intermediate annual reporting is common. When applying for new funding you will normally have to submit a summary of the results of previously funded work, including that which is in progress. Maintaining adequate records makes such reporting much easier.

Box 3.1: NCRIS performance indicators for Infrastructure capabilities

- *Providing Research Infrastructure*
  - ◇ Value of new infrastructure by location
    - + Include cost and description of facilities and equipment
  - ◇ Value of all infrastructure made available under NCRIS by location
    - + Include cost and description
- *Meeting Researcher Needs*
  - ◇ Number, type and location of applicants for each facility
  - ◇ Number, type and location of users for each facility
    - + User types are university, publicly funded research agencies, industry, other
    - + User location in institution
  - ◇ Percentage utilisation of facilities
    - + Based on available capacity
    - + Breakdown per specific capability node and/or instrument if applicable
  - ◇ Measures of user satisfaction
- *Quality of Research Infrastructure*
  - ◇ Benchmark against other Australian and overseas infrastructure, benchmarks may include:
    - + Specific comparisons and instruments or facilities where available
    - + Independent reviews
- *Collaborative Infrastructure Provision*
  - ◇ Extent and duration of collaborative agreements / relationships established for managing and developing research infrastructure
  - ◇ Include type of agreement and parties involved
- *Fostering Collaborative and World Class Research*
  - ◇ Number and nature of Australian research collaborations that involve use of NCRIS infrastructure
    - + Include type of collaborative activity and parties involved
  - ◇ Number and nature of international collaborative research activities supported by NCRIS infrastructure
    - + Include type of collaborative activity and parties involved

---

Specifications from funding source

It is particularly important to follow the requirements from the funding agency with respect to the provision of intermediate and final reports. Failure to do so can result in ineligibility for further funding.

For a simple research grant the reports are likely to comprise a description of the work accomplished with emphasis on outcomes, together with a compilation of outputs such as publications. The format will be specified by the funding agency and needs to be followed carefully. Large programs generally have more complex requirements with detailed reporting against milestones and key performance indicators in addition to the scientific component.

Needs for internal documentation

You should not neglect the benefits to be obtained by preparing effective internal documentation that can supplement the material submitted to the sponsor of the project. Such material will be a major aid when you are preparing publications and presentations. For example, it is worthwhile capturing the *metadata* associated with the work in detail, so that there is a good record of what has been done as well as the results. In addition to common forms of record keeping, such as laboratory notebooks, it is desirable to keep notes on the packages used and maintain documentation of the inputs to software. Another class of information that is often neglected is to make a record of the problems encountered, and the way in which they were tackled. This can be particularly helpful when framing future projects.

In a multi-stranded project, brief reports of progress can be helpful to maintain cross-communication between investigators. These may just consist of an exchange of e-mails, but should be preserved for reference.

Intermediate reporting

Intermediate reports normally represent a concise account of progress, since the previous report. For large programs, reporting is commonly aligned to the milestones and performance indicators agreed at the time of funding, and may include only brief accounts of the scientific endeavour. In all cases one should be realistic about actual progress and not be tempted to make exaggerated claims, which can come back to haunt you. If you are struggling to maintain progress at the rate envisaged, it may be possible to renegotiate targets or at least provide a warning to the sponsor.

Final reports

The character of a final report depends on the style of research and the funding sources. Particularly in the case of commissioned research, you may need to provide full documentation of results and the means by which they were derived. The final report will normally be the reference point for future funding, and so adherence to the requirements of the funding sponsor is critical.

It is worth remembering that the final report may be your main record of a research project, and that you may need to revisit it for your own purposes. The report provides an opportunity to reflect on what has been achieved, but in a busy life may be put together in a hurry in parallel with applications for further funding, so some of the value to you can be lost.

Even where a short formal report is needed, this is a good point to bring together and organise the materials developed in the project. As noted above, such information can substantially aid in the preparation of publications

Interface between financial and scientific reporting

Often the discharge of financial and scientific reports to the same funding agency are undertaken separately. Yet, frequently. no funds settlement is made until the final scientific report is tendered. It is therefore important to ensure that your financial office is aware of the submission of the final report.

3.7 Risk analysis

The risks associated with a project have to be judged in context. Formal risk analysis is now common for large projects, and may be required by institutions before committing to take on a major new research activity of program.

Risk is an inherent part of scientific research, if any new results are to be obtained. Most projects are based on a conjecture about the nature of a problem, which may not be true. Thus, rather than attempt to suppress risk, we need to learn how to manage it. Careful preparation can enhance the probability of success.

3.7.1 Aspects of risk relevant to research programs

We can classify the main types of risk associated with research programs under a group of headings:

Technical risks

- is approach appropriate?
- is necessary expertise available?
- is the timeline feasible?

Infrastructure risks

- dependencies on particular pieces of equipment
- contingencies in event of critical failure

Financial risks

- is costing realistic?
- delivery timing?
- foreign exchange risks?
- withdrawal of support (what are the full range of commitments you are obliged to meet?)
- are there reserves?

Information risks

- availability of critical information
- intellectual property issues
- management of data and metadata

Reputational risks

- what constitutes failure in the project?
- is the institutional reputation likely to be affected by such failure?

3.7.2 Assessment and mitigation of risk

Recognising the existence of risk helps towards the mitigation of its consequences. The checklist we have just presented can be used as a template to run through the issues associated with a particular project, so that the critical issues can be identified.

It is worthwhile to analyse the nature of risks and to provide a rating as high, medium or low. The rating should be based both on the impact of the risk on the project and on the likelihood that the event might be triggered. Typically low-impact scenarios can be ignored, but for high- and medium-impact events a contingency plan should be developed, and included in the risk management toolkit.

Reaction to risk can focus on either taking action to reduce the risk, by reducing its likelihood or impact, or alternatively to control the impact if the event is triggered. The identification of specific risks allows the possibility of planning alternative approaches that avoid this class of event. Contingency planning should proceed far enough that the steps required to secure the most positive result for the project have been identified.

In infrastructure projects it may be possible to adopt some form of *risk transference* so that, e.g., a supplier is required to give a fixed price or guaranteed delivery with the possibility of penalties in the case of noncompliance. Alternatively, it may be appropriate to take out insurance against a risk occurring.

The AuScope national infrastructure project in the Earth Sciences in Australia developed a set of issues which were regarded as key factors

---

Box 3.2: Risk Management Strategy – AuScope

AuScope Ltd will assess and develop risk management strategies in the Interim Implementation Plan, and these strategies will be updated and further developed through successive Annual Business Plans. AuScope will recognise the following areas as key risks requiring management:

- Technical Risks associated with science and engineering issues
- Cost Uncertainties, including
  ◇ Infrastructure and planning risk
  ◇ Foreign exchange risk
- 'Market' Risks, including
  ◇ Demand-related risk
  ◇ Joint venture risk (including membership)
- Physical Infrastructure risk
- Long-term Sustainability risk
  ◇ Including availability of key personnel
- Information security.

---

for risk. The resulting risk management strategy is detailed in Box 3.2. The individual components and institutions involved in this large program were asked to develop assessment of risk against these criteria. They were then required to present their mitigation strategies at the beginning of the funding period, so that these risk-reduction concepts were in place before major investment was made.

Among the most important issues identified were the impact of foreign exchange volatility, the impact of operational costs at the end of the initial phase of funding and the finite lifetime of many pieces of equipment, at least as top-quality facilities.

3.8 Intellectual property issues

The primary goal of research is the creation of new knowledge, which may have potential value in the marketplace. Such value may, in some cases, be exploited explicitly through seeking patent rights. In general the entire set of ways in which a project interacts with prior knowledge and the creation of new knowledge is subsumed under the heading of *intellectual property*.

It can be important to identify the suite of information and understanding that is brought into a project by different project partners. This *background* intellectual property is generally easy to identify, but not necessarily easy to gather together since it may involve know-how that an individual or group brings to the work. This knowledge is not necessarily expressed in any formal way, but may represent cumulated experience that is hard to quantify.

The intellectual property generated by the project may need to be identified in intermediate and final reports. This requires careful consideration of what has been achieved, and whether there are results that are clearly generated by the project that do not depend on either what has been brought into the project in the beginning or drawn in from outside during the course of the work.

When dealing with government agencies and commercial entities, intellectual property issues tend to play a prominent part in agreements and contracts. It is important to understand any restrictions on rights in intellectual property, and legal advice is likely to be needed to make sure that your interests are protected in any formal agreement. It is all too easy to lose control of your own work without adequate compensation.

Chapter 4
Communicating Research

The value of research is lost if it is not communicated to others. The extraordinary insights of Leonardo da Vinci remained locked in his mirror script in his notebooks for centuries, and so did not influence subsequent events in the way that they could have, had they been published.

The scientific paper appearing in a recognised journal remains the primary mode of distributing research results. The form is evolving rapidly in the transition from paper to electronic publishing, but still retains elements from its historical roots. Peer review was introduced for the *Philosophical Transactions* of the Royal Society of London in the 1670s by its first secretary Henry Oldenburg. Such review remains the principal filter on what is published, and the process can be public in some newer open-access journals. The role of journal editors varies significantly between journals, but ultimately they mediate the review comments and make decisions about publication.

The other important way of disseminating research results is through conference presentations. Oral presentations at large meetings are commonly brief – so that care is need to convey the key messages succinctly and effectively. The alternative mode of a poster can provide more space, but again presents challenges for clear communication. A longer seminar gives more scope to develop ideas, but is often given to a more diverse audience so in this case it is important to provide adequate background.

Many projects now have a web presence, and this can provide a useful vehicle to present a broader range of materials. There are many styles employed for web pages, but as with other forms of communication, it is important to take account of the group of people to whom it is targeted. The balance of imagery and text, and the general style can make a significant difference to the appearance and effect of the same material.

The style of scientific communication changes with the years, and it can be useful to look at important papers in the field to analyse how they conveyed their message. Would the same type of approach work today?

4.1 Preparing a publication

The main way in which the results of a research project are communicated to the scientific community is through articles in scientific journals. The article can provide information on the methods and techniques employed, together with information on how the conclusions were reached. In principle, with the aid of the paper, it should be possible for other researchers to follow and reproduce your work. The value of an article is enhanced by good organisation, a good writing style, and clear figures.

4.1.1 Preliminaries

Before you start to develop the structure of a journal article it is worth spending some time on thinking about the aim of the paper, and where is the most appropriate place for it to be published.

Target of paper

A scientific report prepared, for example, as an intermediate product of a project may largely consist of a compilation of the work that has been accomplished, and the details of the way in which it has been carried out. In contrast, a journal article is intended to convey clearly the reason for the work, the way in which the scientific questions have been addressed, and what has been learnt. The paper should aim to build a coherent picture of the endeavour. It is always best to be able to end the article with firm conclusions or results, but do not force the issue to produce strained conclusions that are not substantiated by the earlier parts of the paper.

Not all projects produce the desired positive outcomes, a method may fail in some circumstances or the basic hypothesis may be flawed. Such negative results are of importance for the development of a field, since they help to define the boundaries of successful activity. Yet, it takes some care to present such negative results in a publication in a way that establishes their significance.

Choice of journal

There are huge numbers of scientific journals ranging from the general to the highly specialised. For your publication to have impact it needs to appear in a place where it will be encountered by other interested researchers. The research resources assembled at the beginning of a project should provide a good guide to the range of journals in which relevant work has been published. This list can provide a starting point for your choice. The individual journals tend to develop styles and specialisations that influence the type of material that they receive.

You need to have a good idea of the material that will appear in the paper to be able to assess the appropriate journal to which it should be submitted. Assess your target audience for the results you want to communicate, and then judge whether such people would expect to look at a particular journal. An important test is to ask yourself whether you would you look at this journal? If not, is it the right place for your paper?

Once you have an idea of potential journals it is worthwhile reading a group of articles from each one, so that you get an idea of the style expected for the journal. This will help when you come to craft your own paper.

Sometimes there is pressure to publish in a prestigious general journal such as *Science* or *Nature*. These journals have high rejection rates since they receive far more papers than they can publish. Also the review process and necessary revisions to meet detailed journal specifications can take far more time than in a more specialised journal.

Length requirements

It is often tempting to think that a particular result warrants a short paper, submitted to a specialist letters journal. However, working within a restricted length is not necessarily easier than a longer paper. You need to write concisely and informatively, while making judgements about the specific material that must appear. The choice of figures can take on considerable importance, since there will not be space to devote to a lengthy explanation of a complex figure. Frequently reviewers of short articles will request the inclusion of further details to make the story more complete. It can be hard to balance such clarifications with the length requirements.

A short paper can be an effective way of drawing attention to one aspect of a larger work, and it can be beneficial to work on both in parallel, at least at the beginning. Once the broader context is clear then it can be easier to focus attention on the facet addressed in the shorter work.

Where several authors collaborate on a short paper, stylistic jumps are likely to occur because short segments are being grafted together. It is desirable for one author to go through the whole paper endeavouring to provide a nearly uniform style before submission.

4.1.2 Constructing the manuscript

The process of successful scientific writing is not simple. A paper needs to be interesting and clear if it is to make a significant impact. A crisp and concise style avoiding ornate language is desirable, but too restricted a word usage can create a dull impression.

You are trying to encourage others to take up your ideas, and so you need to make an effort to explain the important points. It can be very effective to pitch your presentation at a level of knowledge slightly lower than you expect from your readers, so that you lead from explanation of the known into the new concepts. It is tempting to make full use of the technical jargon of your field, or employ acronyms to reduce length. Nevertheless, explanations in plain language and variety in description can help encourage the reader to follow your ideas.

Where there are a number of authors then the responsibilities of the members of the team need to be clear from the start, to avoid unnecessary duplication of effort. Many authors will be writing in other than their native tongue, and may be tempted to transfer phrases and constructions with which they are familiar into their writing. Unfortunately, the meanings may not transfer readily and the writing may seem stilted, even if not ungrammatical. Common problems in English come from the use of the definite article 'the' and the indefinite article 'a', particularly for those whose own languages do not employ such articles. It is good practice to get material checked by a native speaker of the relevant language ahead of submission to a journal, this will improve the perception of reviewers and enable the content of the work to be appreciated. It is difficult to assess what is being said when sentence construction is contorted or incomplete, and good work may be turned down because its quality has been obscured by poor language.

Organisation of material

The structure of the paper needs to guide the reader through the important points. However, even well written material may not have the desired impact. It can therefore be beneficial to get a paper read by someone who is not a member of the writing team to see if the logical construction works before the paper is sent to review.

Even experienced authors can be disappointed by how their papers are perceived, since they have got too close to the material and so have a fixed view about how it should be organised. On occasion I have been surprised how a careful presentation may be misinterpreted, or regarded as disjointed.

It is often useful to prepare a skeleton structure of headings as you begin to write a paper. The combination of section and subsection headings allow you to organise the subsequent development. You can add material under the headings to capture the main points you wish to make before writing connected prose. I commonly employ an initial approach with dot points for the main ideas, which then get expanded as the composition proceeds. Figures play an important role in the presentation of ideas and it is useful to have reasonably well-drafted versions early in the writing process so that the overall structure is apparent. As you write you may find that you need to modify the figure to enhance the way in which you express an idea.

Many journals provide templates for manuscript submission for common word processing software, or *LaTeX* used extensively in mathematics and the physical sciences. It is advantageous to use such templates from the beginning of manuscript construction, since the paper will then conform to the requirements of the journal and minimise issues at the time of production.

The style of journal articles is changing with most scientific journals now appearing dominantly in electronic form. This means that more dynamic information, such as movies of simulations or animations can be included, commonly in the supplementary materials. Yet, most users will download the paper as a PDF document to be read offline and only rarely is the supplementary material automatically provided on download.

Title and abstract

This is your chance to grab readers' attention since the title will be the first thing that they encounter and, if sufficiently interesting, they may move onto the abstract. A title should be succinct and informative. Long titles may attempt to capture the full character of the work, but tend to become very cumbersome and lose readers' attention. In general, it would be difficult to reduce as far as the title "Q" used for a study of seismic attenuation by the late Leon Knopoff. Nevertheless, a memorable title may encourage citations.

The abstract is a very important part of the paper, particularly with electronic publication since it is likely to be used to determine whether a paper will be downloaded. The aim is an informative précis of the important aspects of the work and conclusions, not a catalogue of the contents of the article. There are too many examples of abstracts that do not emphasise the principal results and their significance.

At the end of the abstract it is common to include a set of *keywords* that play an important role in both indexing within the journal and the activation of electronic notifications. Often these keywords have to be chosen from a set supplied by the journal, and do not always have as a direct a correspondence to your paper as you would like. Further developments that form part of the evolution of the paper in the electronic age include a 'mini-abstract' in the form of a set of dot point highlights, and visual abstracts incorporating a key figure.

Introduction

The aim of the introduction is to make sure that the reader has a clear idea of where the paper is leading, and how it relates to previous work. There are many different styles employed in different fields. The length of the introductory material expected can also vary considerably. Thus it is important to have a look at a number of papers in a field to get an idea of the style required before getting to far into the writing phase. This is particularly important in interdisciplinary or multidisciplinary work where the journal to which the work is being submitted may have a rather different expectation to those of your home discipline.

The introduction is not just a list of what has already been done. It should provide the motivation for the study and lead into the specific aspects of the work being described. The introduction can also be useful to foreshadow the main outcomes, particularly where these are at variance with earlier work.

Body of text

This is the place where you present your specific approach, results and interpretation. The way in which this is done will vary substantially depending on the character of the field, and the nature of the work. Highly theoretical work will have a distinctly different style to experimental or computational studies.

It is worth noting that a number of journals have developed specific requirements for description of Methods and Data sources, and you need to be aware of such journal requirements as you write. For example, a Methods section may be dispatched to supplementary material for short articles.

Discussion and conclusions

You need to make your ideas count, so this portion of the paper has considerable importance. This part should not be just a reiteration of the interpretation, but an attempt to put it into context and develop the ideas further. The discussion should bring in and expand on issues of the reliability of your results, and the way that they may depend on the assumptions employed, both explicit and implicit from the approach itself. It is legitimate to include some speculative component provided it is clear that this represents ideas for the future.

References

You should follow the journal style as you construct your reference list. There are many formats employed in different journals, and it can become quite cumbersome to adapt material from one convention to another. For example, one may have to convert from a case where initials precede the name for all authors after the first to another in which initials follow, but names are in small capitals. The necessary editing is tedious and prone to introduce errors or omissions.

Many such complications can be reduced by organising a database of references in a standard archive format such as *Refer* or *BibTeX*. The web resources *Web of Science* and *Scopus* that were mentioned in Chapter 1, allow exports of bibliographic information in such standard formats or as text. Appropriate software, such as EndNote, can extract the references from the *Refer* database in the desired form for the journal for use with word processing software. Similarly *BibTeX* can be used with *LaTeX* and a journal class file to give the desired reference structure.

Preparation of figures

Figures represent a compact way of conveying information, particularly for the results of multiple experiments and analyses. Summary diagrams of processes can also be effective. Informative figure captions are important, but do not neglect the need to link to the main text with appropriate explanation so that the content can be appreciated.

Modern computer software offers superior presentation tools than were readily available using pen and ink. However, the quality of figures has not improved as much as might have been hoped. One factor is the temptation to include too much information in a single figure. Another is the use of inappropriate colours; standard yellows disappear against a white page, and red/magenta can often be difficult to distinguish. It is also worth remembering that a significant proportion of men are red/green colour-blind and so may not be able to distinguish lines of different hues. Physiological tests indicate that the eye is much more effective at separating tonal information than colour alone. For map-like plots it is tempting to use the rainbow colour schemes provided as default by much software. Not only does such a scheme suffer from red/green ambiguities, but it is typically uniformly bright. For single-sided information a simple grey tone progression works well, and can be colour tinted to be distinguish between different classes of information. Where both positive and negative values have to be presented, two sets of graded tones away from a neutral colour are very effective.

Sometimes general purpose software is used to generate figures, for example, the use of spreadsheet charts. In general, this approach does not give adequate quality. It is better to transfer the information into a drawing package.

Figures should be created and stored at high resolution. It may be necessary to compress resolution to reduce file size, but retain the high quality original. A particular problem comes from the treatment of lettering when saved as an image. The PNG format preserves the character of fonts better than JPG. Where possible information should be saved in vector form (e.g., EPS or TIFF). You should follow the instructions from the journal, since the publication process may favour certain formats for the figures.

Although it tempting to use a single figure for multiple purposes, you need to bear in mind the different ways that they will be viewed. Larger lettering is needed for effective use on a distant screen presentation, and so figures from presentations are generally not immediately suitable for a journal article. If the original format is available they can be modified, but if only an image is preserved the results are unlikely to be satisfactory.

Supplementary material

In a print journal it was necessary to include all pertinent information in the article. The switch to electronic presentation means that additional information can be included. Indeed reviewers tend now to ask for more information than before because it can be placed in the supplementary material.

The role of the supplementary materials is to add to the main content and enhance the main paper. As far as possible one should avoid duplicating material, but it may be necessary for some explanations to be repeated to make the supplementary material easy to understand. Well chosen movies can be an excellent way of conveying information, but it is helpful to provide information on what is being presented and how this bears on the main arguments of the paper. Because supplementary material is not automatically downloaded with the article it can be irritating to the reader to find it being discussed in the main paper without having immediate access.

Exercise 4-1:
Preparation of a Title and Abstract
Write an abstract of up to 250 words on your current work (or a topic of interest to you). The abstract should have a suitably informative title and provide a succinct account of the most important aspects of your work.
You should choose the most general conference in your field as a target and aim to produce an abstract that can be read by non-specialists.

4.1.3 The editorial process

The dominant mode of submission of material to a scientific journal is now fully electronic via an electronic portal system that assembles the material you have prepared into a package, which can be viewed and assessed by the editors of the journal. Your manuscript and figure material will have to be uploaded and reviewed before the package is completed. This process can take a fair amount of time, depending on the speed of internet linkages or demand on servers. Do not compress figures too far in an effort to speed upload – figure quality is important at the review stage and poor figures are likely to trigger a negative response from reviewers.

In some cases you will be able to suggest the editor whose knowledge of the field makes them the most suitable to handle the material. Frequently though it will be the keywords or classification tags attached to the article during the transmission stage that will determine which editor receives the paper to handle for review.

A number of models of editorial structure are employed by journals. There may be a group of editors who are empowered to make independent decisions, or a principal editor who assigns papers to associate editors who handle the review process. The ultimate decision on the fate of the paper will then usually be made by the principal editor.

You will be frequently be asked for suggestions for suitable reviewers for the paper you have submitted, and indeed for people who you would not wish to review with reasons. The suggestions you make will be taken into consideration by the editors, but are likely to be supplemented by their own perception of the work, and search systems exploiting your keywords or classifiers. You may find it is not easy to make suggestions, nor will it be for the editors. Often a number of people will have to be asked before an adequate number of reviewers can be found. The policy of the journal will determine how many reviewers will be sought.

The continuing increase in the numbers of papers being submitted for publication, places considerable pressures on the editorial process. The editor has to judge between the opinions expressed on a piece of work, and does not always agree with the assessments by the reviewers. In which case they may make an independent judgement or seek a further opinion.

In most cases the editors of scientific journals are working scientists, who fit in their editorial duties around the other demands of their busy lives. Securing reviewers can be a difficult process, and slow response to review is regrettably not uncommon. In consequence, a paper may take much longer to get back to an editor's desk for decision-making than they, or you, would like. It is legitimate to make a query to a journal when the process seems inordinately slow, but harrying an editor is likely to be counterproductive.

A new style of editorial process is beginning to emerge in some open-access journals where the review process is public and the submitted version of the work, the review comments and response are accessible to view. The revised work then appears in the journal of record. Further evolutions in the process can be expected as more journals cease any form of print publication.

4.1.4 Reviewing

The task of the reviewer selected by the editors is to provide a judgement on the quality of the work and presentation. This may well involve completion of a checklist of assessments of different aspects of the manuscripts accompanied by an explanation of the reasons for the recommendation made to the editors. The review report can include confidential comments to the editors as well as material to be transmitted to the authors, which could include an annotated manuscript. Even if you invoke anonymity as a reviewer, it is likely you will leave clues to your identity in the style of your review.

The reviewer will normally be approached by e-mail and need to access the journal system to download the material for assessment. They will then choose how they handle the material, fully electronically or with a hard copy. Lengthy material is likely to get a slow response in any case. As a reviewer you have a obligation to the community to complete your review in a reasonable time. Think about how annoyed you will be for a tardy review of your own work! It is better to decline a review because you are too busy, than drag it out over an excessive time period.

4.1.5 Revision and resubmission

After some time has elapsed, you will receive from the journal a review package normally comprising an editorial comment and the reports from the reviewers. You should take time to digest this material before making any reaction. Indeed you have worked hard on the paper to make it clear, and yet the reviewers have misunderstood your intentions! Despite your efforts you may not have been sufficiently clear, or the reviewers have recognised an aspect of the work that escaped your attention.

Treat the reviews as constructive criticism and work to incorporate as many of the suggestions as you can. You are not required to agree with everything suggested by the reviewers, but will have to justify the way you have undertaken the revision in a response to the reviews, which is submitted with the revised version. Pay particular attention to the most critical comments or suggestions, since these tend to assume greater significance with pressure on journal space.

The revised version will need to be resubmitted through the journal system, with replacement of files. The new version and your response to the reviews will be seen by the responsible editor. In the case of minor, or moderate, revision the editor may choose to accept directly. Where major revision was requested, you can expect that the revised version will re-enter the review cycle. Occasionally new issues emerge with the revision that could have been, but were not, raised with the original version. This is annoying, but you have changed the paper so it will be perceived differently.

4.1.6 Post-acceptance issues

Once the paper has been accepted by the journal it enters the production phase, and the interface switches to the publisher. The first step is the finalisation of the manuscript and figures to incorporate any final minor changes, and queries from the production staff. Then the process of conversion into the electronic forms for archive and presentation begins.

With a variable delay you will receive notification of the availability of electronic proofs, commonly with a request to return in a very short space of time. The proofs will be accompanied by questions raised in the typesetting process, such as ambiguities in the text, and missing or inconsistent references. These questions need to be answered and mistakes in the proofs rectified – remember what the world will see is the product of your checking. A common place for errors to appear is in figure captions, which are often missed in manuscript checks.

Exercise 4-2:
Critique of a recently published paper
Choose a recently published paper in your field, and subject it to your own review.
Does the paper clearly express its goals, and does it achieve them? Are the title and abstract adequately informative? Is all necessary information given to allow the development of the results and the conclusions?
This exercise is valuable for recognising weaknesses in one's own writing.

4.2 Seminars and conference presentations

For all classes of presentation of your scientific work you need to think about the nature of the audience on the particular occasion. The same visual material can be used for different styles of talks with little modification, provided that you adapt the way in which you talk about the science. In all cases you need to make sure that you bring your audience to the point where they can understand your material. Try to pitch the presentation just below the level of preparation that the audience can be expected to have, so that you can carry them with you into the more complex parts of the material.

It takes time and effort to produce a good talk, so do not leave preparation too late. Remember that figures may need to be drawn or at least modified from the form used for publication to achieve the maximum impact in a visual presentation. Images are important in conveying ideas, but do not neglect the value of some words on screen, particularly in reinforcing the message of a figure. In the days when 35 mm slide projection was the norm, dual slide projection was very effective with pictures on one screen and complementary words on the other. The balance is not as easy to achieve with computer-based presentation, but interleaving of text and graphics can be effective. Many conference presentation systems now use a 16:9 aspect ratio for the image on the screen. This configuration can offer the possibility of well balanced parallel figures and text, which tends to be too cramped with the older 4:3 aspect ratio. Do not be tempted to reduce the size of text to include more material on a slide. Can you read the text on the slide when you stand a few metres away from your computer screen? If you cannot, then increase the size and cut down on the text, your audience will appreciate being able to read without straining their eyes.

The visual material seen by your audience should act as a guide to the important points that you want to convey. You should vary your wording from just that on the screen, otherwise you might as well be silent and let people read! In all presentations it is important to package your message and not to try to cram in too much detail. Fewer, well-chosen slides with a limited number of important points well conveyed is much better than information overkill.

Always check your material carefully before you use it for the first time. Spelling errors are particularly annoying to you and your audience; though experience suggests at least one will probably creep in, however careful you have been.

The way in which you present material will be dictated in part by the time available. For a short conference presentation you can only hope to make a few points. In a longer seminar you can be more discursive, but still make sure that your message is clearly packaged.

4.2.1 Short oral presentations

The time slots allocated to oral presentations at major conferences tend to be rather short and so it is critical to make the best possible use of your allocation to convey your message. Clear, uncluttered slides will aid you to capture the essence of your work in an easily digested way. Your talk may come at the end of a session, when the audience is thinking of lunch or the coffee break. You need to be able to gain their attention so that they remember your work in a positive way.

For such a short presentation the timing and balance of material is critical. A practice talk given to a mixed group of specialists and general listeners can help iron out issues with the timing of delivery and clarity. You can get feedback on both content and intelligibility, and acquire some confidence to face a new audience.

It is natural to be somewhat nervous before you give your presentation, but this can be helpful in helping you to project your message on the day. Do not rush – it is better that you present less more clearly than go too fast. If timing of your talk starts goes go wrong (which can happen even with experience), and you are facing the orange light with only a short interval to go, you must know in advance which are the key points that must be made. Simply make these points and stop. You will thus make the best impact you can.

4.2.2 Poster presentation

Poster presentations have become a major component of large scientific meetings, and are preferred by some researchers as being less stressful than giving an oral presentation. They should not, however, be regarded as an easy option. A well designed poster can convey the essence of your message effectively, and may well be able to stand alone without a presenter. But, you need to be present to answer questions and provide further information and can find that you have given the equivalent of several oral presentations during the course of a session.

The instructions for a meeting will include the allowed size of a poster and its orientation. You need to find out how much space you are allowed well ahead of time and design to the specification. All too often people turn up to a meeting with a poster in the wrong configuration that cannot be mounted appropriately.

A good poster has to be well planned so that the results are conveyed clearly and effectively. The material should be kept simple, and well organised by sections with a logical flow, whilst making full use of the space. The style should be consistent throughout. The title is important and should be clearly visible. It is useful to provide an abstract that can be easily read and digested to attract further attention to the work.

The poster needs to be able to be digested from some distance away, particularly if the session becomes crowded. This means the standard text needs to be in a relatively large size (e.g., 24 points or more of a very legible font such as Times). Sans serif fonts are best reserved for headings. Use colour sparingly for emphasis, and avoid the use of prominent and complex background material that distracts from the rest of the poster contents.

When you have finished a draft get it reviewed by others, and see if they grasp the major points that you wish to convey. Avoid public embarrassment by always checking your spelling before you print.

At the session you need to position yourself so as to encourage people to view your work, so do not stand in front or too close to the poster. Allow visitors to digest your work and ask you questions. Even when discussing detail with an interested group try to allow room for the material to be seen by your next customer.

4.2.3 A formal seminar

A full seminar of typically around 50 minutes duration presents a different set of challenges than a short presentation. You will often be speaking at a different institution than your own, and may be in a stressful situation such as a seminar attached to a job interview. You therefore need to understand the nature of your audience and their expectations, and adjust the way in which you talk accordingly.

In all cases, you need to engage your audience and carry them over a considerable time with you as you expound your ideas. Do not be tempted to cram in too much material, a moderate number of slides is adequate. You need to justify to yourself why you need more than one slide every couple of minutes. If you aim for the equivalent of an animated image, then do this explicitly rather than rush on to yet another slide.

Even where you anticipate considerable expertise in the topic among your audience, it is worth providing some general introductory material so that your particular approach can be appreciated by the whole group. It is also useful to take a little time to give a clear exposition of the goals of the work before you start to present the detail. The attention of the audience is sharpest at the beginning of the talk, so it is a good idea to flag the main results early on.

Not only should you explain what you have done, but also why. You will have made choices about, for example, analysis techniques that will influence your results, and you can share your insights with the group. It is worthwhile emphasising the innovative aspects of your work and where they may lead. You should consider breaking up your talk into sections with appropriate lead-in material. The divisions help to provide variety and re-engage the attention of your audience. Bring the talk to close by capturing the main points in a few concluding slides. In this way you can reinforce your message.

Once again, timing is important for a long seminar, and prior practice is useful. It is much better to stop short and leave time for questions than overstay your welcome.

4.3 Project web-pages

An important vehicle for providing outreach for your project is via web pages. Once again you will need to consider what audience you are trying to reach. Generally it will be your scientific colleagues, and so uncomplicated presentations with text supplemented by well-chosen figures with clear explanations can work well.

It is desirable to have a project home page where the goals are outlined and, for larger projects, to have separate pages for the different strands of activity. Make sure that all members of the project group are recognised on the site, so that they can feel ownership.

Often you will have to work within institutional design requirements, and then build on this framework. Where you have a group of related pages it can be advantageous to create pages that share a common structural 'wrapper'. Then insertion of new pages just requires adjustments to the wrapper rather than editing each page individually. Do not neglect to keep the pages up-to-date after they have been created. Out-dated or incomplete material can leave a bad impression.

4.4 Studying classic papers

The styles of scientific communication continue to evolve, and it is useful to look back at classic papers in your field to see how the ideas were presented in their original form. The importance of some critical concepts are recognised immediately, but in other cases the ideas may be so novel or unfamiliar that they take time to influence the broader field. For example, the significance of Albert Einstein's 1905 paper on special relativity became most apparent through the advocacy of Max Planck, who was able to recognise the critical shift in viewpoint associated with the assumption of the invariance of the speed of light.

It can be particularly valuable to undertake a group exercise whereby a number of important papers are discussed, with one individual acting as the reporter for each paper with a short presentation of the key results and a subsequent group critique of the results. Keypoints that can be considered are:

- Does the science still hold?
- Are the arguments convincing?
- Have developments since changed our perception?
- How could the paper have been improved?

Delving back into the literature often reveals that concepts that seem contemporary existed in a recognisable form at an earlier date, but were not exploited because experimental or computational resources did not exist. You will also see that trying to deflect the course of a field is hard, it takes time for entrenched positions to be overturned.

Chapter 5
Research Issues

With the trend towards larger and more collaborative, projects research is becoming a more social activity than hitherto. This means that a broader range of human interactions impinge on projects. In addition to the expectations of a research supervisor, interactions with colleagues and collaborators can play an important role.

Because research pushes beyond the known, even the best laid plans can go awry. Some barriers can be overcome by extra effort, but it is not uncommon for even experienced researchers to reach an impasse. In this chapter we explore what can be done when you are stuck, and how to recognise the symptoms of problems ahead.

Science is built on the principles of full exposition and reproducibility so that work can be tested by others. Even so, there are occasional well-publicised cases of scientific malpractice. We therefore consider the general principles of scientific ethics and try to illustrate the issues that may be faced.

5.1 Working with the research team

In the early stages of a research career, guidance will be provided by research supervisors and advisers. The situation changes once you have your own project, since now you will be put in the position of providing advice and guidance to others.

The success of a project reflects on the whole research team, and it is therefore important that a cooperative attitude prevails. Team engagement is enhanced if there is clarity as to the roles of the different members. Indeed it is very important that you understand the way that all the members of the team view the project, since this could have a major impact on the dynamics of the group and the harmony of working relations. Even if the team is only yourself and a student, it is important that you are both aware of the relative contributions that are expected to be made to the project.

Where a project has multiple strands, communication must be maintained so that components can move together. Even where the same people are involved with different facets of the work, they may not see the interrelations and dependencies. When you have drawn up a project plan it is useful to share it with the team so they understand your expectations. Often they will come up with suggestions that can improve the way that work can be done. In a similar way the team can act as a useful sounding board when progress seems to be slipping, and may be able to provide alternative viewpoints that illuminate the issues.

Early in a project it is important to work out the way in which reporting requirements are handled so that information is available when needed, rather than having to be assembled in a rush at the expense of other activities. Similarly it is useful to map out the expected pattern of presentations at meetings, together with the class of publications to be produced and the anticipated timetable.

5.2 Working with management structures

A funded project brings with it a range of internal and external interfaces to management structures. Just to go through the formalities of accepting a grant or contract requires working with the research support office and possibly legal checking. Subsequently both financial and scientific reports have to be delivered to the external agency, and there can be penalties for noncompliance with agreements.

At the time you start to prepare proposals, it is a very good idea to gain an idea of the structures within your institution with which you will have to deal, and the way that they interact with outside bodies. This will enable you to understand where you can seek advice, and where support may be found with unfamiliar roles. Dealing with contracts gets easier with experience, but is intimidating at first. It is likely that, at the very least, the contract document will be need to be read by lawyers, and not uncommon for complex negotiations to ensue.

Where there are a limited number of available funding schemes, approaching key deadlines place great pressure on research support offices, and it is hard to secure attention for other issues. Needs may therefore have to be flagged well in advance so that appropriate resources are available.

For projects that form part of a larger research program, there will be specific requirements imposed by the central managers of the program. In particular you are likely to be reporting against milestones and key performance indicators, as well as describing the scientific work accomplished. The work is eased if you have a clear project plan, and maintain a regular tracking of progress. Prior reports provide a handy reference, but when editing look out for embarrassing oversights, such as failure to change dates.

5.3 The impact of problems

Few research projects proceed precisely as planned, and indeed not all changes are negative. The work may open up new possibilities that need to be exploited within the agreed structures, or spawn a new proposal and project. Nevertheless, a close guard needs to be maintained so that incipient problems or slippages in progress can be recognised.

In some cases it may be an issue of redirecting resources so that a project component can be put back on track. However, such a shift has to made carefully, since a change in the balance of the components can have unexpected knock-on effects.

If you have made a critical path analysis you will be better placed to judge whether action needs to be taken, or whether some delay can be tolerated. Rapid withdrawal of resources can have a destabilising influence on the research team. If a significant change is proposed you need to make sure that everyone understands why, and should listen to alternative suggestions that may help by viewing the issues from a different angle.

The most difficult situation to handle is where it becomes apparent that the expected research route is not working. In this case you have to put in effort to understand what aspect of the work is causing difficulties, so that you can pose questions and seek advice from other workers in your field before the situation becomes desperate. The more clearly you can express your issues, the more likely it is that you will get helpful responses. Whatever the advice, you need to think about any suggestions carefully, supplementing if appropriate with your own investigations, before making a significant change of direction. It may be that a modest reformulation of the approach will suffice.

A necessary consequence of undertaking research is that you have to accept the possibility that your ideas may be wrong. You are trying to advance knowledge and have to test your conjectures and hypotheses against rigorous tests. A dangerous position is to become so enamoured of an idea that you seek specific information to verify the concept, rather than accept that a broader class of tests might invalidate the idea.

5.4 Ethics in research

Jacob Bronowski in his thought provoking set of essays *Science and Human Values* attempts to define characteristics of the scientific enterprise. Although framed in the 1950s, his characterisation remains the essence of modern science:

> The values of science derive neither from the virtues of its members, nor from the finger-wagging codes of conduct by which every profession reminds itself to be good. They have grown out of the practice of science, because they are the inescapable conditions for its practice. Science is the creation of concepts and their exploration in the facts. It has no other test of the concept than its empirical truth to fact. Truth is the drive at the centre of science; it must have the habit of truth, not as a dogma but as a process. ... Science confronts the work of one man with that of another and grafts each on each; and it cannot survive without justice and

> honour and respect between man and man. Only by these means
> can science pursue its steadfast object, to explore truth.

Bronowski identifies two key principles of scientific endeavour: *Trust* and *Dissent*. Results are accepted until demonstrated to be wrong, and we have to be prepared to have our results overturned by more careful studies or a new leap that supersedes our ideas.

The current perception of the scientific method has been strongly shaped by the view of Karl Popper that hypotheses are to be tested to see if they can be refuted. When tests are adverse, a hypothesis should be discarded and a new development made. In practice the process may be gentler. We still use Newtonian physics for many purposes, because it provides a perfectly adequate description of the behaviour of most macroscopic bodies. Yet, we know that we need a quantum mechanical treatment for the very small, and a GPS navigator requires to take into account the distortions of space-time around the Earth described by general relativity. These more general hypotheses supersede the earlier theory, but as yet have not been subsumed into some more general object. The success of science means that it tends to be somewhat conservative, so that hypotheses are kept in use until the evidence against them becomes irrefutable. As pointed out by Kuhn substantial effort is required to deflect the scientific endeavour, and so change, when it comes, tends to involve a bold redefinition of the prevailing paradigm, which itself remains in place for a substantial period.

To keep scientific standards flourishing and allow the infusion of new ideas requires standards of *ethical* conduct that sustain the honesty and honour of the work, and which are embedded in both methodology and practice. Scientists are expected to respect each other's property, to share appropriately and acknowledge where credit is due; in short *to do unto others as you would wish to be done to yourself*, a sentiment not confined to its presence in many religions.

Many countries have formulated codes of conduct for scientific research that cover both good practice and what needs to be done when the standards are breached. Special care has to be taken where research impinges on human subjects, or experiments are carried out on animals. Well developed procedures require clearance for such activities through review by specific ethics committees, before any work takes place.

The enormous range of scientific activity largely conforms to desirable conduct, but regrettably, there are lapses and such failures to adhere to standards of conduct can be highly publicised. The ethical practice of science reflects broader community values, and underlies trust in scientific results. Yet we see many efforts to discredit scientific results based on selective interpretation with encouragement, on occasion, from commercial

> **Box 5.1:** Examples of research misconduct (after Australian Code for the Responsible Conduct of Research 2007)
>
> There are many ways in which researchers may deviate from the appropriate standards including, but not limited to:
>
> - fabrication of results
> - falsification or misrepresentation of results
> - plagiarism
> - misleading ascription of authorship
> - failure to declare and manage serious conflicts of interest
> - falsification or misrepresentation to obtain funding
> - conducting research without required ethics approval
> - risking the safety of human participants, or the wellbeing of animals or the environment
> - gross or persistent negligence
> - wilful concealment or facilitation of research misconduct by others.

interests. Science has to be strong enough to resist such erosion of the basis of trust, and to stand up for those who espouse unpopular positions based on their analysis of the critical results.

Too often when cases of apparent malpractice emerge it transpires that those involved have pushed their ideas too far. Insufficient care in carrying out the basic work is combined with analysis procedures that are designed to enhance agreement with a strongly held viewpoint. Such self-delusion can rapidly approach or include fraudulent behaviour, for example, by failure to report adverse findings, particularly where commercial interests are involved.

In all their work a research team needs to maintain high standards of intellectual honesty and integrity, with rigour in all scientific and scholarly activities. Due credit must be given to prior work, and conflicts of interests, particularly with respect to intellectual property, managed carefully. It is important that proper practices are followed for safety and security in all environments, and that work is recorded and archived in a responsible manner. Researchers must respect the rights of those affected by their research; approvals from appropriate ethics committees, safety and other regulatory bodies are be obtained as required.

Box 5.1 contains examples of deviations from acceptable behaviour, extracted from the *Australian Code for the Responsible Conduct of Research*. This list provides examples of activities that breach ethical standards, and therefore bring scientific activity into disrepute

An area where considerable contention can arise is the attribution of authorship of a scientific work. Any author must have made a substantial intellectual contribution to the work and so be able to take responsibility for at least part of the total work described. Appropriate contributions include, the conception and design of the project, the analysis or interpretation of the data, or a major contribution to the writing. A number of journals now require a statement of the contribution of authors, and even where not required it is useful to be able to justify the intellectual contributions made

The report *On Being a Scientist: A Guide to Responsible Conduct in Research* prepared by the U.S. National Academies provides a useful resource on ethical issues in science with many illustrations of situations where choices have to be made that have ethical consequences. A broad-ranging discussion of ethics in research is also provided in Chapter 8 of *The Art of Being a Scientist* by Roel Snieder and Ken Larner, who also explore the commercial interface. Both works are very readable and highly recommended; bibliographic details are provided in the suggestions for *Further Reading* at the end of this book.

Chapter 6
Case Studies

In this chapter I describe two scenarios where I have made full use of planning and management tools. The first is a major infrastructure rebuild, where the requirements of the project necessitated full plans. The second is a recently completed project to bring together a wide range of information about the seismological structure of the Australian region to produce a 3-D digital reference model.

The two cases represent different styles of planning, but illustrate the value of using management tools to set the scene, so that effective responses can be made to future eventualities.

6.1 Major infrastructure upgrade

This first case shows how planning tools were used in the redevelopment of the Warramunga Seismic Array in remote outback Australia. The array had been operating since 1976 with a major role in the detection of underground nuclear explosions. In consequence it was designated as one of 50 primary seismic stations in the monitoring network of the 1996 Comprehensive Nuclear-Test-Ban Treaty. The Preparatory Commission for the Treaty (CTBTO) produced a request for a proposal for the physical development of the site, which required a comprehensive plan for the work to be completed within the 1998 calendar year. The array was the first to be upgraded around the world, and so the CTBTO had not yet developed its procedures in full.

The Warramunga array lies in the savannah belt of northern Australia, with two main seasons and occupies a substantial area with stations at up to 25 km from the central recording station. The monsoonal 'wet' occurs in the Austral summer, with a start in late October or November with major electrical storms. The level of rainfall is unpredictable, but the access road to the site may well be flooded at any time after the start of the wet until March or even early April. During this period it is difficult to get heavy equipment to the site. Temperatures rise at the end of the 'dry' season and thunderstorms are common before the rain arrives. These weather considerations dictate the window over which work can be guaranteed.

6.1.1 Specification

With the signing of the 1996 Treaty the entire seismic array has to be refurbished with a complete replacement of all seismometers, data recording and data transmission to bring it up to Treaty standards. In the

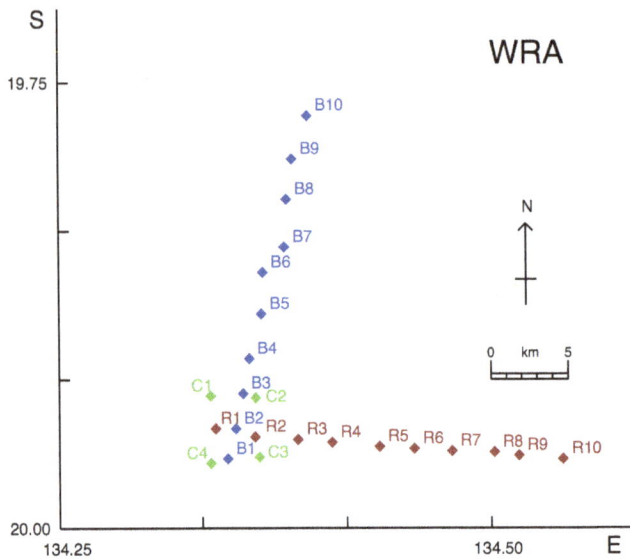

Figure 6.1: Configuration of array, with new stations shown in green.

process, the array is to be enlarged from 20 stations in an L-shape (B., R.) to include a central cluster of 4 new stations (C.), as shown in Figure 6.1. This configuration had been developed following discussions with the CTBTO in Vienna.

The equipment for the upgrade is to be provided by CTBTO, but all site preparation and works have to be organised ahead of installation. The net result is to be an entirely new array, but with the previous geometry included to allow comparisons with prior data.

The start time of the project was specified by CTBTO with the expectation that the work would be completed by the end of 1998. In consequence the planning focused on setting up a time line that would allow completion in October, ahead of the 'wet'. A project plan was required as part of the development proposal, together with a detailed budget in U.S. dollars.

Because this was the first site in a major international investment, there were still uncertainties in the specification of the facilities to be provided by CTBTO at both the central recording facility and the individual sites, since contracts were being arranged in parallel with the site work. The central recording facility is to be maintained in an existing building near site B2 where generator power was available. The data from the remote sites is transmitted by digital radio telemetry to the central recording facility. This meant that line-of-sight communication is required, and the curvature of the earth starts to be important for the trajectory to the outlying elements.

Action	Apr 1	May 1	Jun 1	Jul 1	Aug 1	Sep 1
WORK PROGRAM AGREED						
PROCUREMENT						
TRACK WORK						
PIT REFURBISHMENT						
DRILLING						
CONCRETING						
CENTRAL FACILITIES						
Environmental Monitoring						
FENCING						
SITE PREPARATION COMPLETED						
Central Facilities						
Seismometer installation						
UPGRADE COMPLETED						

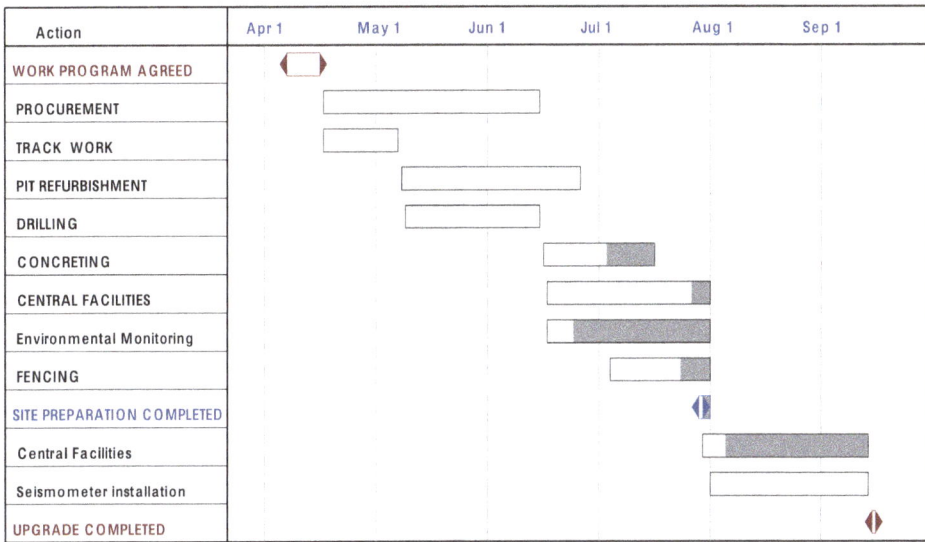

Figure 6.2: Gantt chart – time lines for project elements, with potential leeway shown in darker grey.

At each remote site, new equipment and fencing will need to be installed, with the possibility of drilling to emplace seismometers. Thus heavy equipment has to be able to access each site. Because of the weather conditions this gives a limited window for which such access can be planned. Ahead of the 'wet' season, substantial thunderstorm activity builds up, with lightning likely. Yet, at the same time, the near surface is at its driest and so electrical grounding becomes a major issue.

The team available to carry out the site work comprised the two staff at the array, with extra local help to be hired, and a contract manager in Canberra to secure procurement of necessary goods and services. As mentioned the proposal was required to provide a full project plan, and this proved to be valuable because it identified a number of issues that had not been anticipated in the initial thinking.

6.1.2 Planning

The task was broken into a number of discrete elements associated with the different phases of the work, with time relations dictated by appropriate priorities. Thus, it was necessary to upgrade the tracks before heavy equipment could be brought in to drill new seismometer holes, or concrete pads could be laid at each site. Work at the central facilities could start later than at the outlying sites, but all would have to be ready for the arrival and installation of equipment.

Resource	Action	Apr 1	May 1	Jun 1	Jul 1	Aug 1	Sep 1
Kennett	WORK PROGRAM AGREED						
Davies	PROCUREMENT						
Grant	PROCUREMENT						
Sirotjuk	TRACK WORK						
Sirotjuk	PIT REFURBISHMENT						
Hansen	PIT REFURBISHMENT						
Grant	DRILLING						
Hansen	DRILLING						
Contractor 1	DRILLING						
Contractor 2	CONCRETING						
Grant	CENTRAL FACILITIES						
Sirotjuk	CENTRAL FACILITIES						
Hansen	CENTRAL FACILITIES						
Davies	CENTRAL FACILITIES						
Sirotjuk	Environmental Monitoring						
Hansen	Environmental Monitoring						
Contractor 3	FENCING						
Grant	Central Facilities						
Supplier	Central Facilities						
Grant	Seismometer installation						
Sirotjuk	Seismometer installation						
Hansen	Seismometer installation						
Supplier	Seismometer installation						
Kennett	UPGRADE COMPLETED						

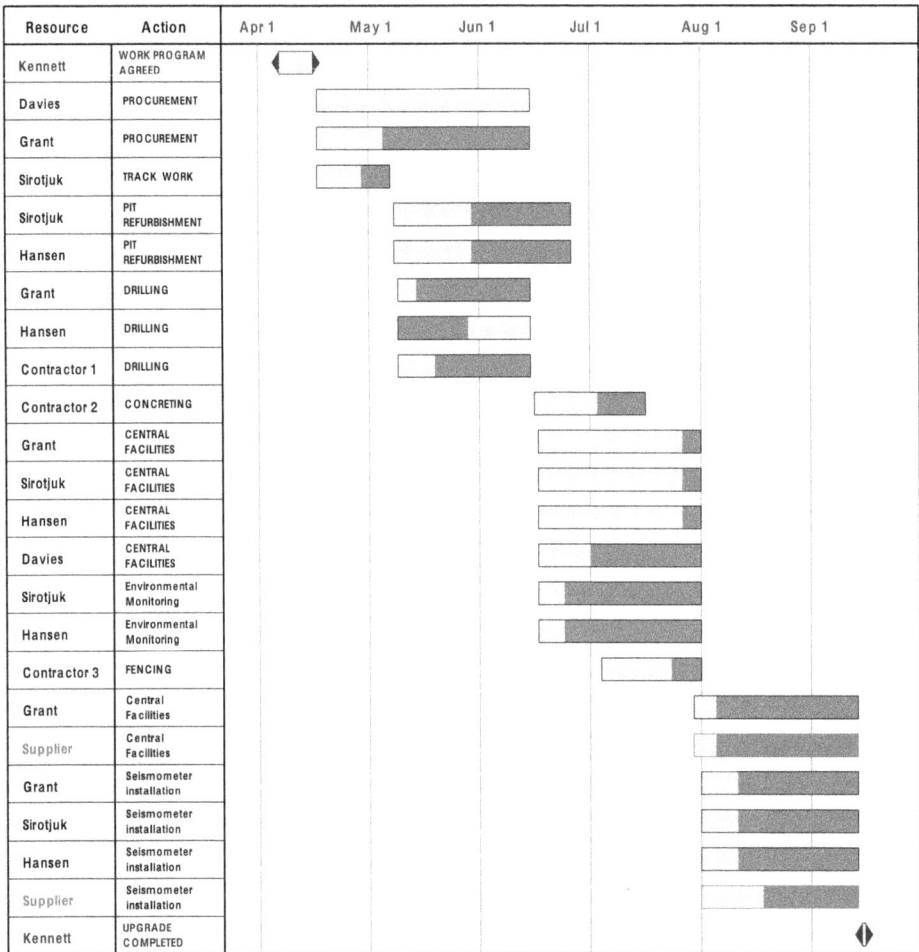

Figure 6.3: Personnel deployment for project elements, with multiple people used to move forward the different aspects of the projects.

The Gantt chart and personnel resources chart for the project plan are shown in Figures 6.2 and 6.3. The grey areas indicate where some leeway was allowed for the project component.

The analysis indicated that drilling would need to start as soon as practicable, but we could not engage a drilling contractor until the contract was signed with CTBTO. Nevertheless we were able to establish the earliest availability of drilling. The next issue was the critical role played by the concrete work. To stay on an October completion schedule, less than one day could be allowed at each site. This meant that the contractor would have to enlarge his team so that form work and concrete poring could proceed in parallel.

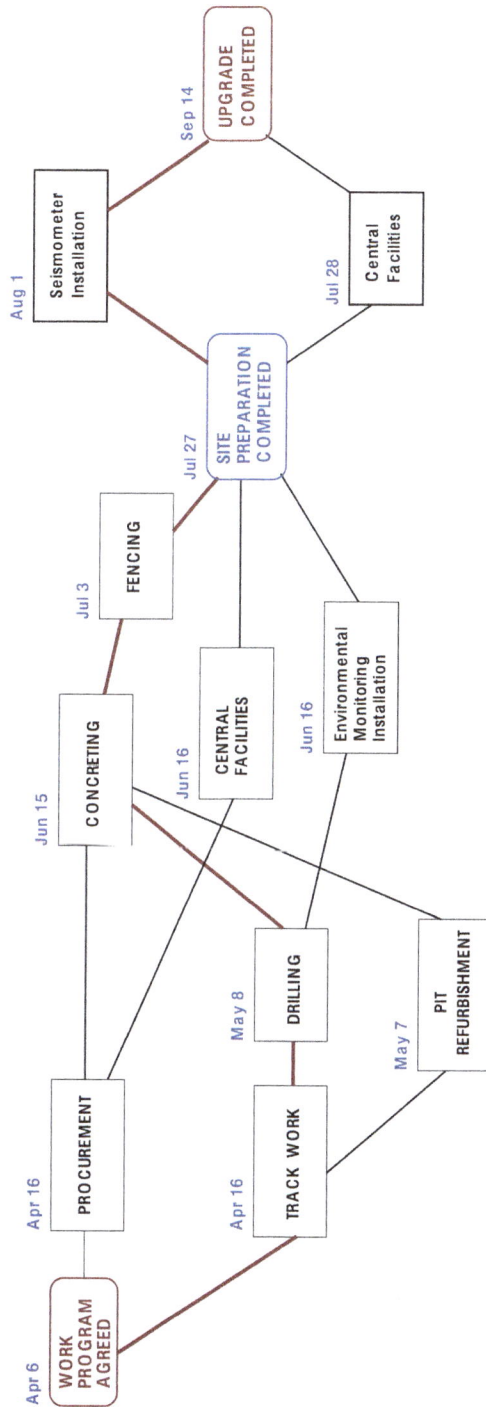

Figure 6.4: Critical path analysis for upgrade project.

The critical path analysis is shown in Figure 6.4, and it became very clear that it would be difficult to have everything ready by the beginning of August so that equipment from CTBTO could be installed. By design, the completion data was set at mid-September to allow the possibility of some slippage and still stay ahead of the 'wet'.

6.1.3 WRA project in practice

The project plan was accepted by CTBTO, but then there were substantial delays in processing of the proposal and budget approval. An immediate consequence was that the time line would have to be started later, but this then impacted on weather conditions. This was a period of volatile exchange rates and we had some anxieties about the costing in U.S. dollars when expenditure was in Australian dollars.

One of the problems that emerged when the project had been approved to begin, was that the original specifications from CTBTO, included in the request for proposal, had been changed once the actual contracts for the equipment were drawn up. In particular the data acquisition units for the remote sites had double the original power consumption. This required both a new concept for housing to reduce the effect of external heating from the sun (day-time temperatures can easily reach 40+°C in summer), and increased battery capacity required for the solar power system. Both changes increased the costs per site.

The original concept for the central data receiver tower was too vulnerable to wind, and was replaced by a 21 m self-supporting structure raised by professional riggers. Thanks to the high central tower, line of sight to the far stations was achieved with 12 m masts at these sites.

The originally proposed time frame for preparing the concrete slab, mast footings and fencing at each of the seismometer sites was too tight, but this was accommodated in the revised plan. The delays in approval caused complications in the availability of the drilling contractors needed for the new sites and mast footings, but fortunately the weather held and the work was completed before any rain came.

The delivery times for the site masts and batteries was much longer than anticipated, so the original timetable could not have been met. However, the delays meant that the design and build of data acquisition housing was feasible because of suspension of work over the 'wet'.

The actual completion date was in September the following year, a full year later than originally indicated. However, CTBTO equipment was not available before July of the second year so the original required time frame was unreasonable.

Fortunately, the project plan had correctly identified the elements of work to be done, and the necessary sequence, and provided a good basis for organization. Sufficient contingency funding had been included so that unexpected costs could be absorbed. Careful bulk purchasing and tight control kept costs in check.

6.1.4 Lessons from project

A clear project plan was required as part of the proposal, but this proved to be a valuable tool. The recognition of the dependencies between the components of the project at an early stage helped to eliminate bottlenecks. Delays to aspects of the project due to external factors can be dealt with more easily when linked to a project plan. The most complex issue was changes in the specifications of the requirements during the course of project. This arose because of the separation of site preparation and hardware procurement.

For this project, undertaken at a remote location, clear communications were important. The contract manager in Canberra was in regular communication with the staff on site, and kept track of the procurements and overall project status. He reported to me as project manager, and together we were able to update the project plans and monitor expenditure against the budget. Visits to the site each year helped us to understand the situation, to rectify small problems, and consolidate the team working on the project.

Up to this point I had used informal planning for projects. I was initially disconcerted to be required to develop formal plans, but rapidly appreciated their value in forcing detailed thinking about the project, and providing a clear framework for management. I have since made substantial use of such approaches.

6.2 Analysis of a research project

The second case study is for the Australian Seismological Reference Model (AuSREM), a project to develop a 3-D model of crustal and mantle structure beneath the Australian region. The model was designed to build on existing work, integrating a wide range of different classes of data and depending on a broad network of collaborators.

The project was first suggested in April 2010 to provide a showcase of work on Australia that could be presented at two major international meetings, which were being held in Australia in the following two years: the General Assembly of the International Union of Geodesy and Geophysics (IUGG) in Melbourne in July 2011, and the International Geological Congress (IGC) in Brisbane at the end of July 2012. Funding

Table 6.1 Basic structure of AuSREM project.

Topic	Begin date	End date
Model Definition	8/11/10	26/11/10
Data Collection	8/11/10	23/06/11
Data Assembly	3/01/11	29/11/11
Tomography	07/02/11	21/07/11
Model Construction	04/04/11	03/04/12
Model Delivery	17/01/11	19/11/12
Completion	09/10/12	14/12/12
Milestones		
IUGG Melbourne	27/06/11	
IGC Brisbane	30/07/12	
End of Project	20/12/12	

was sought to be able to have the reference model presented in full at the IGC meeting in 2012, with publication as soon after as was practicable.

Joint investment from the AuScope infrastructure project and the Australian National University provided support for a research associate for two years and some travel support for collaborators. A recent project funded by the European Union provided a useful background, and some confidence that the work could be completed in a two-year time frame.

The research associate was able to start work on the project in November 2010, and then the first task was to make sure that we had a clear definition of the nature of the model and the way in which it would be constructed. Basic plans had been needed for the project proposal, but the concept of the nature of the work had evolved somewhat before the major effort got underway. The model would comprise crustal and mantle components with a horizontal resolution of 0.5 degrees in latitude and longitude, and be grid based so that interpolation would be able to provide physical property values (e.g., seismic wavespeed) at any point in three-dimensions.

The AuSREM model was intended to build extensively on existing data sources and models, which required the collection of information from a wide variety of sources and subsequent assembly into new products. I was fortunate in securing strong support from a number of collaborators who had worked on different aspects of the structure beneath Australia, and the challenge was to bring all the available information together and capture it in digital form. The clear deadline for presentation in July 2012 meant that the model had to exist some time before, and this guided the overall time-line for the different aspects of the project.

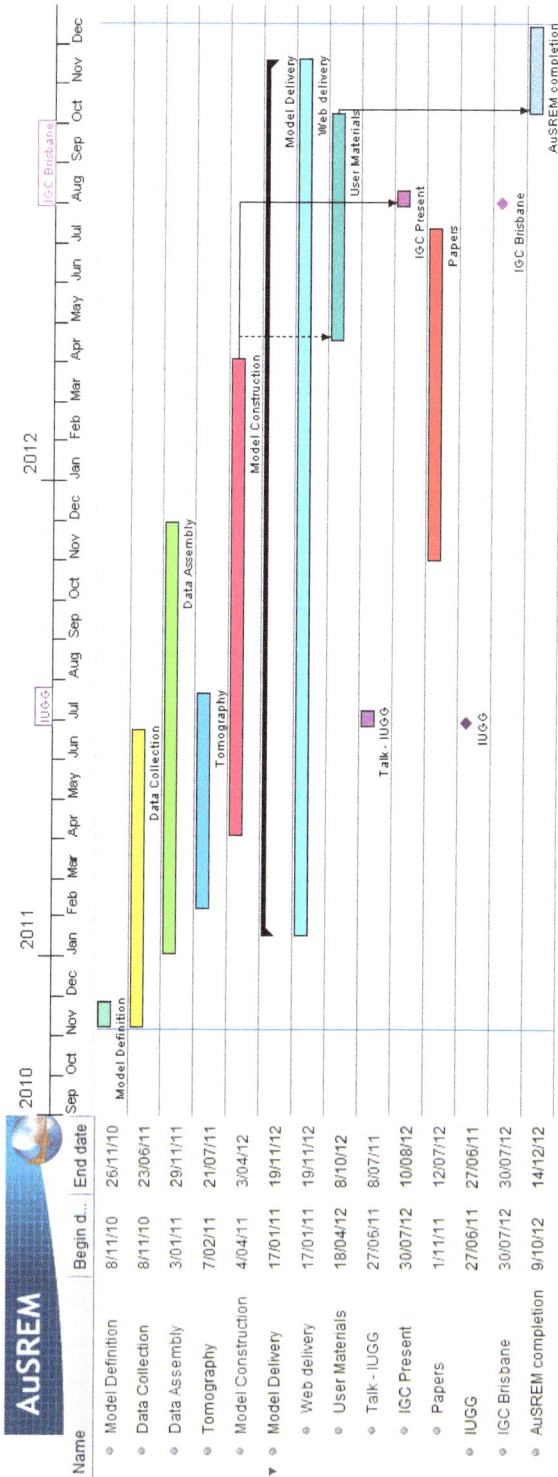

Figure 6.5: Basic structure of AuSREM project with milestones.

The general structure is shown in Table 6.1, and in the form of a Gantt chart in Figure 6.5. From early in the project we established a website where information on the project was provided and products could be made available as they were completed. Model delivery thus formed a constant theme in the project.

An important decision in the model definition process was to decide to build a model for the Australian crust separate from that for the mantle beneath, with connection through a representation of the Mohorovičić discontinuity (Moho) at the base of the crust. The research associate took primary responsibility for the crust and I looked after the Moho and mantle, but there was considerable feedback between the components.

For the crust a new model was constructed using a wider range of information than had hitherto been available. This required considerable collection and assembly of data of different types, followed by the merger of the various classes of information into a suitable model.

For the mantle a different approach was taken, with the model being built on the foundation of a number of exiting studies. It proved possible to get the authors of a number of recent models for the Australian mantle together just before the IUGG meeting in Melbourne. The discussions illuminated the needs, and the final AuSREM mantle model was able to build on updates made by the participants.

From the beginning a detailed project plan was developed using full project tools, which were used to identify the necessary sub-components. Figure 6.6 shows the final development of the project. The earlier versions started with approximately the same start dates for the different elements, but used rather shorter durations until it became clearer how much work was required.

The interaction between project elements is more complex than indicated in Figure 6.6, and some aspects of the work do not fall neatly into a single box. Nevertheless the Gantt chart provides an idea of the complexity of the total effort, and the consequent need to keep track of progress so that the milestones could be met.

In the lead up to the presentation at the IGC in Brisbane, considerable effort was put in to assembling a group of papers to describe the model and its construction. An overall summary paper was prepared for the *Australian Journal of Earth Sciences*, and detailed papers on the crust and mantle model provide details of the techniques used. A new model for the depth to Moho was published separately in late 2011. The publication dates of the materials are indicated by displaced blocks added to the basic Gantt chart.

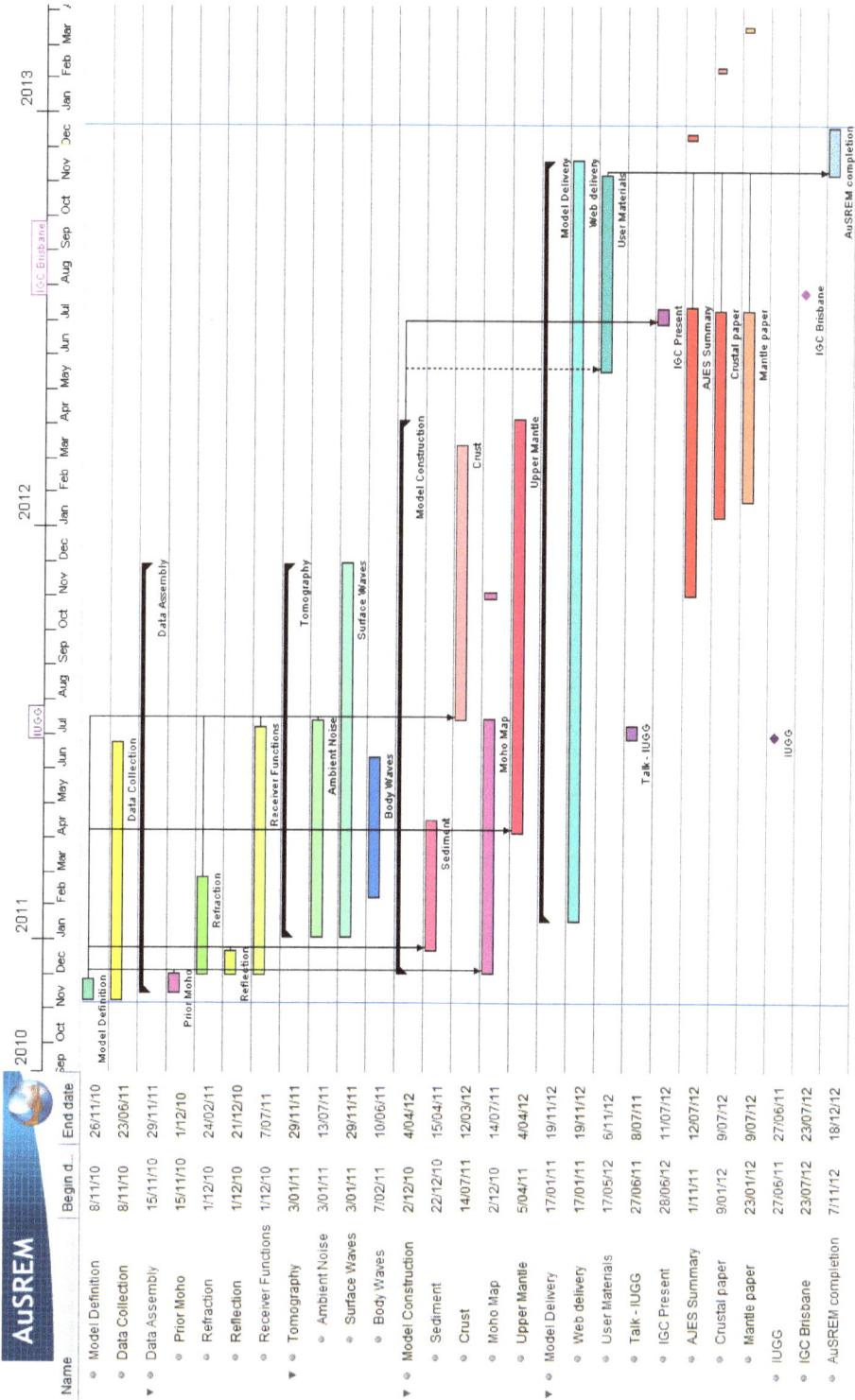

Figure 6.6: Detailed structure of AuSREM project, including sub-components.

The presentation of the results and the preparation of the publications did not constitute the end of the work. Considerable effort remained to set up suitable ways for the model to be used, via interactive visual displays as well as model downloads. We had fortunately recognised the class of commitment that would be needed, and so were able to achieve full delivery in the two-year funded span.

In this case the use of the project tools was of considerable importance early on, because it forced attention on the various facets of the work that needed to be addressed, and the way that they interacted. The initial draft concentrated on the identification of the elements, and then the time relations were refined. The plan was then an important tool for tracking progress and its existence played a major role in the success and timeliness of the entire project.

6.3 General considerations

Starting to prepare a detailed project plan forces you to think carefully about a broad range of issues. Nearly always some features or interrelations emerge that were not apparent in advance. With a modest investment of time in a plan, but consequent concentrated thought, you will have a clearer idea of how the desired outcomes can be achieved in a timely manner. This is already a major return on your effort.

Both of the projects used in the case studies involved interaction with a large group of people. The project plans helped with communication of needs and expectations so that there were few surprises, even when delivery slippages required adjustment.

When you are able to communicate your goals and the way in which you propose to achieve them clearly, your team is able to respond effectively. Their thoughts can be important and can reveal aspects of the project that you have not considered. Do not be dismissive, but see whether you can improve the plan and its management.

Further Reading

R. Snieder & K. Larner (2009) *The Art of Being a Scientist, A Guide for Graduate Students and their Mentors,* Cambridge University Press.

A useful book that is designed principally for students starting graduate research in science, but which also picks up many issues relevant to mentoring and advising research. The work provides an extensive reference list. The perspectives on scientific communication and ethics are particularly useful.

J. Bronowski (1956) *Science and Human Values.*

Originally published in the *Journal of Higher Education Research*, these three essays have been brought together and published in a number of forms. They provide a provocative insight into the creative process, the nature of scientific enquiry and its ethical background.

Committee on Science, Engineering, and Public Policy, National Academy of Sciences, National Academy of Engineering, and Institute of Medicine (2009) *On Being a Scientist: A Guide to Responsible Conduct in Research,* Third Edition. National Academies Press, Washington D.C, U.S.A.

This work and associated audio and visual material available from the National Academies Press website, http://www.nap.edu, provide a useful exploration of situations where ethical considerations come to the fore.

Australian Government (2007) *Australian Code for the Responsible Conduct of Research,* Prepared by the National Health and Medical Research Council, the Australian Research Council and Universities Australia.

This code of conduct, which is freely available on the web, provides an exemplar of the kinds of considerations that define responsible practice and what has to be done when ethical standards are breached. The discussion of *authorship* provides a clear set of guidelines, with some useful examples of circumstances that do not justify inclusion in the author list.

www.ingramcontent.com/pod-product-compliance
Lightning Source LLC
Chambersburg PA
CBHW061222270326
41927CB00022B/3467